"It's amazing how many toys are based upon physics and chemistry principles. Learning science concepts with toys is an exciting adventure for children. Their natural interest and curiosity in science combined with their desire to 'play' with toys provides great motivation to learn."

Jeannie Tuschl—Tulip Grove School, Nashville, Tennessee

"I really learned that there are many toys that can be used to teach science. I hope to expand the use of the TOYS concept in my classroom."

Elizabeth Henline—Mount Orab Middle School, Mount Orab, Ohio

"The toys, experiments, and activities are classroom-friendly to students of all ages."

Mary Hurst—McKinley Elementary School, Middletown, Ohio

"TOYS is a great program, with lots and lots of new and exciting ideas to use in the classroom."

JoAnne Lewis—Stanberry Elementary, King City, Missouri

"I would highly recommend TOYS for all science teachers."

Sarah Birdwell—Butterfield Junior High, Van Buren, Arkansas

"With TOYS, science really becomes part of everyday experiences and materials."

Mary White—Monmouth High School, Monmouth, Illinois

"Teaching Science with TOYS is a wonderful way to motivate children. It's a super program!"

Cindy Waltershausen—Western Illinois University, Monmouth, Illinois

"I received so many new ideas to try out in my classroom that my students will be enjoying learning science without even realizing it!"

Regina Bonamico—Chauncy Rose Middle School, Terre Haute, Indiana

"TOYS activities will enable me to help develop a love for science, genuine inquiry, and higher-level thinking skills with my students. TOYS provides a wealth of ideas to introduce hands-on learning."

Rita Glavan—St. Pius X, Pickerington, Ohio

Teaching Science with TOYS Program Staff

K–3 Team

Dwight Portman
Physics Teacher
Winton Woods High School
Cincinnati, Ohio

Mickey Sarquis
Assistant Professor of Chemistry
Miami University Middletown
Middletown, Ohio

Mark Beck
Science Specialist
Indian Meadows Primary School
Ft. Wayne, Indiana

4–6 Team

Beverley Taylor
Associate Professor of Physics
Miami University Hamilton
Hamilton, Ohio

John Williams
Associate Professor of Chemistry
Miami University Hamilton
Hamilton, Ohio

Cheryl Vajda
Teacher
Stewart Elementary
Oxford, Ohio

7–9 Team

Jim Poth
Professor of Physics
Miami University
Oxford, Ohio

Jerry Sarquis
Professor of Chemistry
Miami University
Oxford, Ohio

Gary Lovely
Physics Teacher
Edgewood Middle School
Hamilton, Ohio

Tom Runyan
Science Teacher
Garfield Alternative School
Middletown, Ohio

Other Staff

Lynn Hogue
TOYS Program Manager
Miami University Middletown
Middletown, Ohio

Susan Gertz
Document Production Manager
Miami University Middletown
Middletown, Ohio

Teaching Physics with TOYS

Activities for Grades K–9

Beverley A.P. Taylor

James Poth

Dwight J. Portman

Terrific Science Press
Miami University Middletown
Middletown, Ohio

**LEARNING
TRIANGLE
PRESS**
▲

*Connecting kids, parents, and teachers
through learning*

An imprint of McGraw-Hill

McGraw-Hill

A Division of The McGraw·Hill Companies

Terrific Science Press
Miami University Middletown
4200 East University Blvd.
Middletown, Ohio 45042
513/424-4444

pbk   9 10 MAL/MAL 0 3 2

This material is based upon work supported by the National Science Foundation under Grant Number TPE-9055448. This project was supported, in part, by the National Science Foundation. Any opinions, findings, and conclusions or recommendations expressed in this material are those of the authors and do not necessarily reflect the views of the National Science Foundation.

Library of Congress Cataloging-in-Publication Data

Teaching physics with toys : activities for grades K–9 / by Terrific
    Science Press, Beverley A.P. Taylor, James Poth, and Dwight J. Portman
        p. cm.
    Includes index.
    ISBN 0-07-064721-6 (pbk.)
    1. Physics—Study and teaching—(Elementary)—Activity programs.
    2. Physics—Study and teaching—Secondary—Activity programs.
QC39.5.T33 1995
372.3'5—dc20   95-2022
                                                        CIP

Acquisitions editor: Kimberly Tabor

Contents

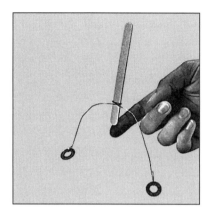

Activities for Grades K–3 9

Activities for Grades 4–6 93

Acknowledgments

The authors wish to thank the following individuals who have contributed to the success of the Teaching Science with TOYS program and to the development of the activities in this book.

Activity Developers/Peer Writers:

Mark Beck	Indian Meadows Primary School	Ft. Wayne, IN
Alison Dowd	Talawanda Middle School	Oxford, OH
Sally Drabenstott	Sacred Heart School	Fairfield, OH
Lynn Hogue	Terrific Science Programs	Middletown, OH
Anita Kroger	Hamilton County Office of Education	Cincinnati, OH
Jo Parkey	Smith Middle School	Vandalia, OH

Teacher Mentors:

Mark Beck	Indian Meadows Primary School	Ft. Wayne, IN
Gary Lovely	Edgewood Middle School	Hamilton, OH
Tom Runyan	Garfield Alternative School	Middletown, OH
Cheryl Vajda	Stewart Elementary School	Oxford, OH
Sue Walpole	FERMCO	Cincinnati, OH

Terrific Science Press Design and Production Team:

Susan Gertz, Amy Stander, Lisa Taylor, Thomas Nackid, Stephen Gentle, Anne Munson

Reviewers:

Alexander Dickison	Seminole Community College	Sanford, FL
Wallace Freeman	Indiana University of Pennsylvania	Indiana, PA
Sandra Harpole	Mississippi State University	Mississippi State, MI
H.T. Hudson	University of Houston	Houston, TX
Jerry Meisner	University of North Carolina at Greensboro	Greensboro, NC
Raymond Turner	Clemson University	Clemson, SC
Sandy Yorka	Denison University	Granville, OH

University and District Affiliates:

Matt Arthur	Ashland University	Ashland, OH
Zexia Barnes	Morehead State University	Morehead, KY
Sue Anne Berger	Colorado School of Mines	Golden, CO
J. Hoyt Bowers	Wayland Baptist University	Plainview, TX
Joanne Bowers	Plainview High School	Plainview, TX
Herb Bryce	Seattle Central Community College	Seattle, WA
David Christensen	The University of Northern Iowa	Cedar Falls, IA
Laura Daly	Texas Christian University	Fort Worth, TX
Mary Beth Dove	Butler Elementary School	Butler, OH
Dianne Epp	East High School	Lincoln, NE
Babu George	Sacred Heart University	Fairfield, CT
James Golen	University of Massachusetts	North Dartmouth, MA
Richard Hansgen	Bluffton College	Bluffton, OH
Cindy Johnston	Lebanon Valley College of Pennsylvania	Annville, PA
Karen Levitt	University of Pittsburgh	Pittsburgh, PA
Donald Murad	University of Toledo	Perrysburg, OH
Hasker Nelson	African-American Math Science Coalition	Cincinnati, OH
Judy Ng	James Madison High School	Vienna, VA
Larry Peck	Texas A & M University	College Station, TX
Carol Stearns	Princeton University	Princeton, NJ
Doris Warren	Houston Baptist University	Houston, TX
Richard Willis	Kennebunk High School	Kennebunk, ME
Steven Wright	University of Wisconsin–Stevens Point	Stevens Point, WI

Foreword

As a child or as an adult, most of us find it difficult to walk past a colorful display of toys without pausing, smiling, and taking a closer look. The urge to roll the truck down the hill, bounce the Silly Putty™, or wind up the walking dinosaur is nearly irresistible. We typically associate toys with fun, discovery, and creativity. In contrast, if presented with a display of physics and chemistry experiments, "fun, discovery, and creativity" unfortunately would not be the words that come to most peoples' minds.

In 1986, a group of colleagues at Miami University wanted to give teachers (and through teachers, students) the opportunity to find out that "fun, discovery, and creativity" are words that very much describe the exploration of physics and chemistry principles. Our idea was to teach basic physics and chemistry principles using toys, thus capitalizing on the natural attractiveness of toys and also showing that physical science is an integral part of our everyday experiences.

We were fortunate to obtain funding from the Ohio Board of Regents to offer the first Teaching Science with TOYS course for teachers here at Miami University. During that year we worked with a tremendous group of teachers who took the toy-based physical science they learned back to their classrooms with wonderful results. Through funding from the National Science Foundation, we have been able to continue the TOYS program here at Miami University and around the country through our University and District Affiliates. Over the past eight years, we have met and worked with over 700 educators from across the United States and around the world. During these years we have continued to develop and test new toy-based science activities in our courses. This book is a product of those years of development.

The teachers we have worked with in the TOYS program are as different from each other as any group of people can be. They come from urban areas, rural settings, and everywhere in between. Some had lots of science background; others had none. Their students reflect the diversity of our nation's schools. But despite their differences, these teachers have had at least one thing in common—the desire to give each of their students the chance to enjoy learning about physical science. Through their feedback in the laboratory and via classroom testing, each of these dedicated professionals has contributed to this book. We thank them.

From TOYS participants and from our follow-up evaluations, we know that toy-based science is making a difference in classrooms, schools, and entire districts. We believe that the best way to learn to teach discovery-based science is to experience it yourself with support from experienced colleagues, just like the best way for your students to learn science is to do discovery-based science with your support and guidance. However, we realize that all teachers do not have the opportunity to attend professional development courses such as the one we offer. Thus, we are delighted to be able to offer many of the activities from our program in book form. We welcome you to join TOYS teachers around the country through your use of the activities in this book and wish you all the best in your teaching.

Teaching Science with TOYS Staff

Introduction

Science is asking questions about the world we live in and trying to find the answers. Students of science should do more than memorize definitions and parrot facts; to make sense of science, students must be given opportunities to make connections between scientific phenomena and their own world.

With this book, you and your students will delve into the mysteries of such fun toys as Push-n-Go® and Ping-Pong™ Puffer. You will learn why come-back toys come back and study the principles behind balancing toys and boomerangs. Your students will experience the minds-on, hands-on learning that will improve their problem-solving skills while also improving their science content knowledge.

Teaching Science with TOYS

The National Science Foundation-funded Teaching Science with TOYS project is located at Miami University in Ohio. The goal of the project is to enhance teachers' knowledge of physics and chemistry and to encourage activity-based, discovery-oriented science instruction. The TOYS project promotes toys as an ideal mechanism for science instruction because they are an everyday part of the students' world and carry a user-friendly message. Through TOYS and its affiliated programs, thousands of teachers nationwide have brought activity-based science into their classrooms, using teacher-tested TOYS activities. Through written materials such as this book, many more teachers and students can share in the fun and learning of the TOYS project.

Becoming Involved with the TOYS Project

By using the activities in this book, you are joining a national network of teachers and other science educators who are committed to integrating toy-based science into their curricula. If you have a TOYS Affiliate (a college or university science educator who conducts local TOYS programs) in your area, you may want to get involved in local TOYS programming.

Another way to extend your TOYS involvement is by attending a Teaching Science with TOYS graduate-credit course for teachers of grades K–12 at Miami University in Ohio. The program uses a workshop-style format in which participating teachers receive instruction in small, grade-level-specific groups. Teachers do not need a special science background to participate in the program; the program faculty review relevant science principles as they are applied to the toy-based activities being featured. Participants spend much of their time exploring hands-on, toy-based science activities.

TOYS is open to all science educators of grades K–12. Applicants should be actively teaching or assigned to a science support position. For more information or an application, write Teaching Science with TOYS, Miami University Middletown, 4200 East University Blvd., Middletown, Ohio 45042; call 513/424-4444 x421; or e-mail *hoguelm@muohio.edu.*

Using this Book

Teaching Physics with TOYS—Activities for Grades K–9 is a collection of some of the physics activities used in the Teaching Science with TOYS professional development program for teachers at Miami University. The TOYS activities have been compiled into this and the companion *Teaching Chemistry with TOYS—Activities for Grades K–9* as a resource for teachers who want to use toy-based physical science activities in the classroom, but who may not be able to attend a TOYS workshop at the Miami site or one of the Affiliate sites nationwide. The activities do not assume any particular prior knowledge of physical science and complete activity explanations are included.

Organization of the Book

This book consists of three main sections: the introductory material, the toy-based physics activities, and the appendixes (an index by key process skills, an index by topics, and an alphabetical listing of activities).

The TOYS activities in this book are divided into three grade-level groupings: K–3, 4–6, and 7–9. Each activity provides complete instructions for use in your classroom. These activities have been classroom-tested by teachers like yourself and have been demonstrated to be practical, safe, and effective in the targeted grade-level range. Each activity includes a photograph of the toy or activity setup and the following sections:

- Grade Levels: Although the activities are specifically written for a particular grade level, many activities can be used successfully by teachers of younger or older students. Suggested grade levels for the science activity and cross-curricular integration are listed.

- Key Science Topics: Key science topics are listed for all activities.

- Student Background: If appropriate, background knowledge or experience is suggested that would be valuable for students prior to doing the activity.

- Key Process Skills: These process skills assist teachers whose lesson planning includes a process focus. See Appendix A for an index of key process skills for all activities.

- Time Required: An estimated time for doing the Procedure is listed. The time does not include Extensions or Variations. This estimate is based on feedback from classroom testing, but your time may vary depending on your classroom and teaching style.

- Materials: Materials are listed for the Procedure and for any Extensions or Variations. Materials are divided into amounts per class, per group, and per student. Most materials can be purchased from grocery, discount department, or hardware stores. Quantities or sizes may be listed in English measure, metric measure, or both, depending on what is clear and appropriate in each case.

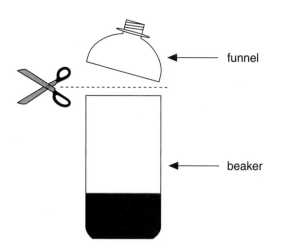

Measuring containers for liquids can be made by pouring a measured amount of a liquid into a disposable cup and marking the height of the liquid with a permanent marker. Beakers and funnels can be made from empty 1-liter or 2-liter plastic soft-drink bottles as shown to the left. Sources are listed for unusual items.

- Safety and Disposal: Special safety and/or disposal procedures are listed if required. See pages 7–8 for a more detailed discussion of safety and disposal issues.

- Getting Ready: Information is provided in Getting Ready when preparation is needed before beginning the activity with the students.

- Procedure: The steps in the Procedure are directed toward you, the teacher, and include cautions and suggestions where appropriate. Quantities or sizes may be listed in English measure, metric measure, or both, depending on what is clear and appropriate in each case.

- Variations and Extensions: Variations are alternative methods for doing the Procedure. Extensions are methods for furthering student understanding of topics.

- Explanation: The Explanation is written to you, the teacher, and is intended to be modified for students.

- Cross-Curricular Integration: Cross-Curricular Integration offers ideas for integrating the science activity with other areas of the curriculum.

- Further Reading: Further Reading provides suggested readings for teachers and students to extend their knowledge of the topic.

- Contributors: Individuals, including Teaching Science with TOYS graduates, who contributed significantly to the development of the activity are listed.

- Handout Master: Masters for handouts are provided for some activities. These may include data, assessment, and/or observation sheets as well as other types. Permission is granted to copy these for classroom use.

Notes and safety cautions are included in activities as needed and are indicated by the following icons and type style:

Notes are preceded by an arrow and appear in italics.

Cautions are preceded by an exclamation point and appear in italics.

Pedagogical Strategies

TOYS program staff members recognize that no single instructional strategy best meets the needs of all students at all times. Howard Gardner, author of *The Unschooled Mind*, views each learner as possessing a distinctive profile of "intelligences" or ways of learning, remembering, performing, and understanding. These differences are thought to dramatically affect what instructional approach is most likely to be effective for a given student. For example, kinesthetic demonstrations, which involve students in role-playing and dramatic simulations, can be useful in providing an understanding of the particle nature of matter and other relatively abstract science concepts.

With this in mind, a variety of instructional approaches can be effectively used to present the toy-based science activities in this book. We have included some suggestions for incorporating guided and open-ended inquiry, process skills and cross-curricular integration of science, learning cycles, and cooperative learning. These pedagogical strategies are based on modern methods of science education that originated with theories of cognitive functioning introduced by Jean Piaget (Piaget, 1958); these theories can be summarized by the following statements:

- If students are to give up misconceptions about science, they must have an opportunity to actively reconstruct their world view based on exploration, interpretation, and organization of new ideas (Bybee, 1990).
- Exploration, interpretation, and organization of new ideas are most effective in a curriculum where hands-on, inquiry-based activities are integrated into learning-cycle units (Renner, 1988).
- Hands-on, inquiry-based activities that involve conceptual change, problem-solving, divergent thinking, and creativity are particularly effective in cooperative learning situations (Johnson, 1990).

The general suggestions in this pedagogical strategies section are provided by teachers who have used TOYS activities in a variety of ways. Some teachers have used a selected series of activities from this book as the basis of a special classroom unit on toys in science which lasted for several weeks. Other teachers have used individual activities throughout the school year in conjunction with the topics ordinarily taught. This section is intended to provide you with some ideas for integrating the TOYS activities into your own curriculum. These TOYS activities are not intended to comprise a stand-alone curriculum; rather, the activities are intended to complement and enrich your own curriculum. We encourage you to consider how these activities can meet the needs of your students and fit your own teaching style.

The Particle Nature of Matter

To be able to explain the nature and interactions of matter, it is necessary to know about the nature of the particles that make it up. As adults, you have probably heard and used the terms atoms, molecules, and ions. Remember, however, that young children do not have the conceptual background to correctly distinguish between these three terms. In the elementary grades, we strongly recommend that you use the general term "particles" with students to prevent creating misconceptions that will later be difficult—if not impossible—to correct. Additionally, with elementary students, limit your discussion to the general idea that all matter is made up of particles that are too small to be seen by the human eye and that, in spite of our inability to see these particles, their existence accounts for matter as we know it. The discussion of the various types of particles (atoms, molecules, ions) should be held to later years in the curriculum when the students are conceptually ready for this level of detail. Our recommendations are consistent with those of the American Association for the Advancement of Science *Benchmarks for Science Literacy* and the National Science Foundation-funded S_2C_2 (School Science Curriculum Conference) report, a joint report of the American Chemical Society and the American Association of Physics Teachers.

Webbing and Charting

In introducing any new unit, you may wish to begin by stating the general topic and asking the students to help you identify some more specific subtopics. With the general topics and subtopics identified, students are then ready to consider three important questions: What do we already know about this topic? What do we want to find out about this topic? How can we find out?

Questions should be considered in this order, with students being encouraged to provide many answers to each. Record the responses on a large easel-sized chart so that the responses can remain on the class bulletin board throughout the unit for continual reference. We recommend that you have your students address one question at a time and that you allow sufficient time for discussion and response before the next question is addressed. Be sure that students think of an answer to "How can we find out?" for each item listed in "What do we want to find out?" In addition to acting as facilitator of this lesson, you may wish to act as recorder and list the student responses on the charts. By first jotting down cryptic notes from the students' responses and subsequently recopying the notes into neat handwriting and complete sentences, the teacher can model an important part of the writing process often not shown to students.

Process Skills and Cross-Curricular Integration

The skills of science and other subjects such as language arts and math are remarkably similar. Process skills such as sequencing, recording information, communicating, and classifying span many disciplines. Primary students can become motivated readers and writers through science activities. Older students develop their abilities to identify and control variables, make meaningful conclusions, and communicate ideas clearly. Most of the activities in this book have specific suggestions for cross-curricular integration. Appendix A lists the principal science process skills that are employed in each of the activities.

As the teacher, it is important that you avoid the common trap of being a forecaster of what to wait for or what will happen. Rather, enable your students to be involved in the process of science and to construct their own understanding. Challenge them to make predictions. Help them record their predictions, do the experiment, record their observations, and reflect on how their observations and predictions varied or were the same. By using constructivist methods, you can help your students develop the skills necessary to eventually take responsibility for their own learning. Provide your students with multiple examples of a given concept to help them develop a foundation necessary for reliable predictions. Encourage them to become involved in the activity.

Challenge your students to be good predictors and good communicators. Young students whose writing skills are limited can still predict and communicate through speaking and drawing. Science affords opportunities to expand students' vocabulary. It affords opportunities for recording procedures so that the activity can be repeated at home. It affords the students a common ground for success with immediate reinforcement of accomplishing a task at hand. Science provides motivation to take on difficult challenges while helping younger children develop the concentration skills necessary to complete multiple-step activities.

Learning Cycles

You can select groups of activities from this book to create a variety of learning-cycle units. A learning cycle consists of a three-phase approach to actively involve students in investigative science experiences. Each phase plays a distinct role in enabling students to challenge their old views about the unit's topic so they can successfully reconstruct and internalize new science concepts. The three phases of the learning cycle each have specific objectives and are presented in order. Different authors may have different names for the phases, but the phases always play the same basic roles. Here, we call them exploration, concept introduction, and concept extension. An awareness activity can precede the exploration phase, but is not absolutely necessary. However, omitting any of the three phases or presenting the phases out of order reduces the effectiveness of the learning cycle.

- The awareness activity uses a thought-provoking or discrepant event to help students realize that they may have preconceived ideas about a particular science topic. Students will be more receptive learners if they become aware of their own preconceptions.
- The exploration phase challenges students' preconceptions through a hands-on activity in which they are told how to make observations and collect data but are not given any new vocabulary or explanations about what to expect. The teacher assumes the role of facilitator, posing questions and assisting students as they work.
- The concept introduction phase asks students to form new ideas based on their observations. The teacher helps students see how patterns in data reveal the concepts being defined in the lesson. The teacher may further develop concepts using textbooks, audiovisual aids, and other materials.
- The concept extension phase allows students to extend their understanding of the topic by using what they know to solve a new problem or conduct an experiment.

Cooperative Learning

For late primary or intermediate students and older, you may wish to use some formal cooperative strategies while doing these activities. Some activities already contain specific cooperative-learning suggestions. Cooperative learning has been well documented as enhancing student achievement. Several popular models for cooperative groups have been described by D.W. and R.T. Johnson, Robert Slavin, Spencer Kajan, Eliot Aronsen, and others. Although these models vary, they typically include elements of group goals or positive interdependence and individual accountability. Often, cooperative models suggest specific roles for each student. If students have not been using cooperative learning routinely, time must be spent at the beginning of the assignment explaining individual accountability and group interdependence and reviewing social skills needed for cooperative group work. Students should understand that their grades are dependent upon each person carrying out the assigned task. The teacher should observe the groups at work and assist them with social and academic skills when necessary.

Safety

Hands-on activities and demonstrations add fun and excitement to science education at any level. However, even the simplest activity can become dangerous when the proper safety precautions are ignored, when done incorrectly, or when performed by students without proper supervision. The activities in this book have been extensively reviewed by hundreds of classroom teachers of grades K–12 and by university scientists. We have done all we can to assure the safety of the activities. It is up to you to assure their safe execution!

Be Careful—and Have Fun!

- Never attempt an activity if you are unfamiliar or uncomfortable with the procedures or materials involved. Consult a high school or college science teacher for advice or ask them to perform the activity for your class. They are often delighted to help.

- Activities should be undertaken only at the recommended grade levels and only with adult supervision.

- Always practice activities yourself before performing them with your class. This is the only way to become thoroughly familiar with an activity, and familiarity will help prevent potentially hazardous (or merely embarrassing) mishaps. In addition, you may find variations that will make the activity more meaningful to your students.

- Read each activity carefully and observe all safety precautions and disposal procedures.

- You, your assistants, and any students observing at close range must wear safety goggles if indicated in the activity and at any other time you deem necessary.

- Special safety instructions are not given for everyday classroom materials being used in a typical manner. Use common sense when working with hot, sharp, or breakable objects, such as flames, scissors, or glassware. Keep tables or desks covered to avoid stains. Keep spills cleaned up to avoid falls.

- Recycling/reuse instructions are not given for everyday materials. We encourage you to reuse and recycle the materials according to local recycling procedures.

- In some activities, potentially hazardous items such as power tools are to be used by the teacher only to make a toy or set up the activity. These items appear under the heading "For Getting Ready Only."

- Read and follow the American Chemical Society Minimum Safety Guidelines for Chemical Demonstrations on the next page. Remember that you are a role model for your students— your attention to safety will help them develop good safety habits while assuring that everyone has fun with these activities.

- Collect and read the Materials Safety Data Sheets (MSDS) for all of the chemicals used in your experiments. MSDS's provide physical property data, toxicity information, and handling and disposal specifications for chemicals. They can be obtained upon request from manufacturers and distributors of these chemicals. In fact, MSDS's are often shipped with the chemicals when they are ordered. These should be collected and made available to students, faculty, or parents should anyone want MSDS information about specific chemicals in these activities.

ACS Minimum Safety Guidelines for Chemical Demonstrations

This section outlines safety procedures that must be followed at all times.

Chemical Demonstrators Must:

1. know the properties of the chemicals and the chemical reactions involved in all demonstrations presented.

2. comply with all local rules and regulations.

3. wear appropriate eye protection for all chemical demonstrations.

4. warn the members of the audience to cover their ears whenever a loud noise is anticipated.

5. plan the demonstration so that harmful quantities of noxious gases (e.g., NO_2, SO_2, H_2S) do not enter the local air supply.

6. provide safety shield protection wherever there is the slightest possibility that a container, its fragments or its contents could be propelled with sufficient force to cause personal injury.

7. arrange to have a fire extinguisher at hand whenever the slightest possibility for fire exists.

8. not taste or encourage spectators to taste any non-food substance.

9. not use demonstrations in which parts of the human body are placed in danger (such as placing dry ice in the mouth or dipping hands into liquid nitrogen).

10. not use "open" containers of volatile, toxic substances (e.g., benzene, CCl_4, CS_2, formaldehyde) without adequate ventilation as provided by fume hoods.

11. provide written procedure, hazard, and disposal information for each demonstration whenever the audience is encouraged to repeat the demonstration.

12. arrange for appropriate waste containers for and subsequent disposal of materials harmful to the environment.

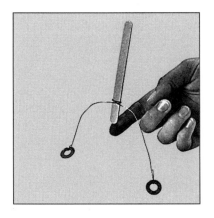

Activities for Grades K–3

Gravity Makes Things Fall

Comparing Mass Using a Pan Balance

Measuring Mass Using a Pan Balance

Ramps and Cars

Balloon on a String

Ping-Pong™ Puffer

Balancing Stick

The Skyhook

The Six-Cent Top

Bouncing Balls

Snap, Crackle, Pop

Magic Balloon

Magnet Cars

School Box Guitar

GRAVITY MAKES THINGS FALL

We know that objects fall when dropped. But why? Students explore the effects of gravity.

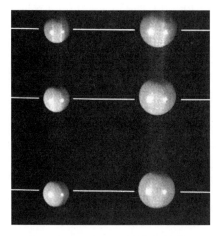

Stop-action photo of falling balls

Materials

For the "Procedure"
Part A, per student
- any small object that does not bounce and is safe to be dropped by a young child (for example, a ball of clay or Play-Doh® or a bean bag)

Part B, per class
- access to a sink with water

Part C, per class
- bathroom scale
- blocks, toy cars, or chairs

Safety and Disposal

No special safety or disposal procedures are required.

Introducing the Activity

Illustrate pushes and pulls as forces; for example, a pushing force is needed to move a baby stroller or a grocery cart, a pulling force is required to move a wagon. Let students push or pull different objects like blocks or cars or chairs around the classroom. Also, try some forces that don't produce motion, like pushing on a wall.

Procedure

Part A: Dropped Objects

1. Give each student an object to drop.

2. Have students drop their objects.

3. Ask the following questions: "Where did it fall? Will it fall to the ground next time you drop it?"

4. Have children trade objects and repeat Step 2.

5. Ask students, "Is someone or something pushing or pulling on the object to make it fall to the ground?"

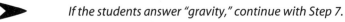 *If the students answer "gravity," continue with Step 7.*

6. If the students answer "no," explain that an invisible force called gravity is causing their objects to fall.

7. Explain that the force of gravity causes all objects to fall and keeps all objects on the ground.

8. Point out to the students that when they push a chair, they are producing the force on the chair. Ask, "What produces the force of gravity?"

9. (optional) If you plan to include any of NASA's *Toys in Space* activities in your curriculum, now would be a good time to ask the students, "What would happen to the objects if gravity were not present?"

Toys in Space: Exploring with the Astronauts, *by Carolyn Sumner is available from TAB/ McGraw-Hill. Toys in Space II toy kits are available from TEDCO, (800) 654-6357.*

Part B: Water Falls Down a Drain

1. Half-fill the sink with water.

2. Have the students watch the water go down the drain. (See Figure 1.)

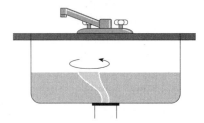

Figure 1: Half-fill a sink with water and observe the vortex that forms as the water drains.

3. Have students put their hand in the water near the drain. They can see the water being pulled down the drain and feel the force by placing their fingers in the swirling water. Tell the students that they are watching gravity at work.

Part C: Weight on a Bathroom Scale

1. Have student volunteers stand on a bathroom scale.

2. Explain that the weight displayed on the scale is a measure of how strong the force of gravity is between themselves and the Earth.

Explanation

The following explanation is intended for the teacher's information. Modify the explanation for students as required.

Gravity is a force that acts at a distance, unlike a "contact" force that we might use to push a chair across the floor. In 1661 Isaac Newton formulated the Universal Law of Gravitation, which states that two objects do not need to be in contact to exert forces on one another. Newton's explanation of gravity states that two objects attract each other with a force that depends on how much mass they possess and the distance between them. The greater the masses the greater the force of attraction. The greater the distance between them the weaker the force.

More formally, Newton's Universal Law of Gravitation states that every object attracts every other object with a force that is directly proportional to the mass of each object and inversely proportional to the square of the distance between them.

When an object falls to the ground, it does so because a force of attraction exists between the mass of the object and the mass of the Earth. They pull each other. The combined masses of the Earth and the object give rise to the force of attraction. We usually don't think of the object contributing to the force of gravity, because when an object (such as a ball of clay) is dropped, the ball is the only object that moves because the Earth is so much more massive than the ball of clay. (See Figure 2.)

Figure 2: When an object (such as a ball of clay) is dropped, it moves toward the Earth.

Cross-Curricular Integration

Art:
• Have students draw pictures with the title, "Gravity at Work."

Language arts:
- Read aloud or suggest that students read one or more of the following books:
  - *Chicken Little,* retold by Steven Kellogg (W. Morrow, ISBN 0-688-07095-0)
    Chicken Little and his feathered friends are alarmed that the sky appears to be falling and that they will become easy prey to sly Foxy Loxy.
  - *Gravity Is a Mystery,* by Franklyn M. Branley (Harper & Row, ISBN 0-06-445057-0)
    This first- and second-grade book explains gravity in simple pictures and words.

References

Croft, D. *Activities Handbook for Teachers of Young Children,* 5th ed.; Houghton-Mifflin: Boston, 1990.

Hewitt, P.G. *Conceptual Physics,* 5th ed.; Little, Brown: Boston, 1985; pp 125–142.

Contributor

Sally Drabenstott (activity developer), Sacred Heart School, Fairfield, OH; Teaching Science with TOYS peer writer.

COMPARING MASS USING A PAN BALANCE

Does a bigger object always have more mass than a smaller object? Does the shape of an object affect its mass? Do two objects always have more mass than one?

Using a balance to compare weights

GRADE LEVELS

Science activity appropriate for grades K–3
Cross-Curricular Integration intended for grades K–3

KEY SCIENCE TOPICS

- mass

STUDENT BACKGROUND

It is helpful if students have had some experience with pan balances.

KEY PROCESS SKILLS

- observing — Students compare the masses of objects using pan balances.

- hypothesizing — Students form hypotheses about the mass of an object when the size or shape is changed.

TIME REQUIRED

Setup   15 minutes
Performance 30 minutes
Cleanup   10 minutes

Materials

For the "Procedure"
Part A, per group of 3–4 students
- pan balance, made for primary students
- 1 large empty box such as a cereal box
- 1 small empty box such as an individual-serving cereal box
- 1 scoop such as a coffee scoop

Parts B and C, per group
- 4 balls of clay or Play-Doh® with the same mass
- pan balance with pans
- 1 container, such as a coffee can, holding enough sand to fill small cereal box

 Instead of sand, you can use any relatively dense material that comes in small enough units to fit into the small box (such as marbles or BBs).

Safety and Disposal

No special safety or disposal procedures are required.

Introducing the Activity

1. Introduce the pan balance and its use—to compare the masses of objects. Show its parts.

2. Demonstrate that if a more massive object is on one pan and a less massive object is on the other pan, the pan with the more massive object will go down and the one with the less massive object will go up. Also show students that the pans will balance if the two objects are equal in mass.

Procedure

Part A: Do larger objects always have more mass than smaller ones?

1. Have students look inside the two cereal boxes to observe that they are empty. Ask students, "Which box has more mass?"

2. Have students check their predictions by placing each box on one pan of the balance. Ask students, "What happened?" *The larger box goes down, so it must have more mass.*

3. Instruct student to use the scoop to fill the smaller box with sand. Ask again, "Which box has more mass?" Answers may vary. Ask students, "How can we check our predictions?" Place the boxes on the pans as before. The small cereal box now makes the balance go down, showing that it has more mass. (In this example, the smaller box contains more material (matter) than the larger box, so the smaller box has more mass than the larger box.)

Part B: Do two objects always have more mass than one?

1. Tell a student in each group to hold a ball of clay or Play-Doh® in each hand. Ask students, "Do these balls of clay have the same mass? How can we be sure?"

2. Have students in each group place one ball of clay in each pan of the balance. Ask students, "What happened?" *The pans balance, so both balls must have the same mass.*

3. Have the students in each group take one of the balls of clay and make it into four smaller balls. Ask, "Now, which has more mass—this one ball of clay or these four balls of clay?" Let students check their predictions.

4. Ask, "What happened?" *The pans balance. There are more pieces on one pan, but not more mass.*

Part C: Does the mass of an object change when the shape of the object changes?

1. Tell a student in each group to hold one ball of clay in each hand. Both balls should be the same size. Ask students, "Do these balls of clay have the same mass? How can we be sure?"

2. Have students in each group place one ball of clay in each pan of the balance. Ask students, "What has happened?" *The pans balance, so both balls must have the same mass.*

3. Instruct a student in each group to roll one ball into a long snake, without breaking it. Ask, "Now, which has more mass—this ball of clay or this long snake of clay?" Let students check their predictions.

4. Ask, "What happened?" *The pans balance. The ball and the snake must have the same mass. Changing the shape did not change the amount of matter in the ball of clay and therefore did not change its mass.*

Explanation

 The following explanation is intended for the teacher's information. Modify the explanation for students as required.

Young students do not distinguish between size and mass. Early experiences often lead to the conclusion that big things have more mass than small things. This may or may not be true.

In addition, young students have difficulty in understanding the concept of conserving mass. If the appearance of the object is changed in some way (broken into pieces or changed in shape), students often assume the mass is changed. Working through these activities provides examples that contradict these misconceptions.

Assessment

Divide the class into small groups and give each member of each group an identical sheet of paper. Have each student fold his or her sheet into a shape different from everyone else's in the group. Have each group make a prediction about which piece of paper has the most mass and which has the least mass. Then have the groups compare the masses of the folded papers on the pan balance by placing one paper on each side. Have the groups explain their findings.

Math:

- Study the concepts of "greater than" and "less than." Introduce the signs < and >, or review. Draw pictures of an alligator's head using the < or > as the main line of the open mouth (See Figure 1), in between two bags or boxes containing different amounts of something. (This could be two bags containing different amounts of marbles, two bags containing different amounts of cookies, etc.) Make sure the alligator's mouth is facing the larger quantity. (The hungry alligator always wants to eat the largest amount possible.)

Figure 1: The hungry alligator always wants to eat the largest quantity.

Reference

Davidson, A. *Understanding Mathematics—Five Teachers' Resource Book;* Shortland: New Zealand, 1984; pp 107–109.

Contributors

Sally Drabenstott (activity developer), Sacred Heart, Fairfield, OH; Teaching Science with TOYS peer writer.

Janet Kinsella, Worth County R-III Elementary School, Grant City, MO; Teaching Science with TOYS, 1993.

Jill Stelzer, Murlin Heights Elementary School, Dayton, OH; Teaching Science with TOYS, 1993–94.

Peggy Tomlin, Queen Anne Elementary School, Lebanon, OR; Teaching Science with TOYS, 1993.

MEASURING MASS USING A PAN BALANCE

How many bears will balance a bag of sand?
Students use counting bears and a pan balance to compare and measure the mass of objects.

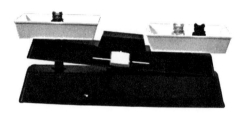

Using a pan balance to measure mass

GRADE LEVELS

Science activity appropriate for grades K–3
Cross-Curricular Integration intended for grades K–3

KEY SCIENCE TOPICS

- mass
- standard units of measure

STUDENT BACKGROUND

It is helpful if students have had some experience with pan balances. This activity should be completed after doing the Teaching Science with TOYS activity "Comparing Mass Using a Pan Balance."

KEY PROCESS SKILLS

- estimating — Students estimate how many standard units are needed to balance a bag of sand.

- measuring — Students determine the masses of different objects using pan balances.

TIME REQUIRED

Setup	30	minutes
Performance	35–45	minutes
Cleanup	10	minutes

Materials

For "Introducing the Activity"
Per class
- (optional) flannel board
- (optional) assortment of pictures of objects that are too massive to lift, such as a car, house, horse, airplane, elephant
- (optional) assortment of pictures of objects that students could lift, such as pencil, dish, apple, cereal box

For the "Procedure"

Per learning center or group of 2–4 students

- 1 pan balance
- 6 small zipper-type plastic bags
- ¼ cup of each of the following:
  - sand
  - rice
  - puffed rice
  - unpopped corn
  - popped corn
  - small buttons
  - 10–20 counting bears

Counting bears are available from Cuisenaire, (800) 237-0338.

  - measuring cup
  - (optional) bear stickers or a bear stamp

For the "Extension"

Per class

- seesaw
- laundry or other type of basket
- rope
- 14 plastic 2-L soft-drink bottles or 9 1-gallon plastic bottles

Safety and Disposal

No special safety or disposal procedures are required.

Getting Ready

1. Fill zipper-type plastic bags for each group with approximately ¼ cup of each of the following: sand, rice, puffed rice, unpopped corn, popped corn, and buttons.

2. Determine the mass of each bag ahead of time using counting bears as units of mass. Adjust the amount of material in the bag so that it balances with a whole number of bears.

Introducing the Activity

Options:

- Introduce the terms "more massive" and "less massive." One way to do this is to do a class flannel board activity. Divide a flannel board into two sections. Have an assortment of pictures of objects that are too massive to lift: car, house, horse, airplane. Have some pictures of objects that students could lift: pencil, dish, apple, cereal box. Have students take turns placing pictures on the correct side of the flannel board. Emphasize the terms "more massive" and "less massive."

- Introduce the pan balance and its use—to measure the mass of objects. Look at its parts.

- Demonstrate that if a more massive object is on one pan of a pan balance and a less massive object is on the other pan, the pan with the more massive object will go down and the one with the less massive object will go up. Also show students that the pans will balance if the two objects are equal in mass. Use counting bears in the following ratios to illustrate these concepts:

  ○ two bears in the pan versus one bear in the pan,

  ○ two bears in the pan versus two bears in the pan, and

  ○ two bears in the pan versus no bears in the pan.

Procedure

Part A: Comparing the Masses of Two Materials

1. Have the students pick up a zipper-type plastic bag of sand in one hand and a zipper-type plastic bag of rice in the other hand. Ask students: "Which bag feels less massive? Which bag feels more massive?" Wait for answers. "Let's compare the masses of the bags on your pan balance to see if your answers are correct."

2. Have the students place one bag on each side of the pan balance. Ask, "What happened?" *The bag of sand went down and the bag of rice went up.* Ask, "Which has more mass? Which has less mass?" *The bag of sand has more mass and the bag of rice has less mass.*

3. Have the students repeat Steps 1 and 2 with different combinations of dry materials.

Part B: Measuring the Masses of Two Materials

1. Review the fact that equal masses on the pans make the pans balance.

2. Have each group place the bag of sand on one pan. Ask, "What happened?" *The bag went down.* Ask, "How can we make this pan come up even with the other pan?" *Add something to the other side.*

3. Ask, "How many counting bears do we need on this pan to balance the sand?" Accept answers. "Let's see which estimate is correct."

4. Have each group place the bears—one at a time—on the empty pan until the pans balance. Ask them to count the number of bears and record this number on the "Measuring Mass with Bears" Data Sheet (provided).

5. Have students repeat Steps 2–4, replacing the bag of sand with the bags of other materials.

6. Have students study the completed chart and determine which material is the least and which is the most massive. Have students rank the materials from least to most massive and arrange the bags from least to most massive.

7. Ask students to suggest other objects that could be used as a standard measure of mass.

Extension

If your playground has a seesaw, use plastic 2-L soft-drink bottles or gallon jugs full of water to determine the masses of student volunteers as follows: Tie a laundry basket to one end of a seesaw. Have a volunteer sit at the other end of the seesaw. (Make sure that the basket and the volunteer are equal distances from the center of the seesaw.) Place bottles of water in the basket one at a time until the seesaw balances or is as close to balancing as possible.

Since body size can be a sensitive subject, even to children, we strongly recommend using a few student volunteers for this Extension. The quantities of bottles suggested in "Materials" are enough to determine the mass of a person who weighs 70 pounds.

Explanation

The following explanation is intended for the teacher's information. Modify the explanation for students as required.

Mass is a measure of the quantity of matter in an object. To measure mass, you determine the number of standard units that will balance the object. In the "Procedure," you use counting bears as a standard unit of measure. In the Extension, you use bottles of water.

If you never had to communicate information about mass to anyone else, any standard unit of measure would be all right. Counting bears or bottles of water would be as good as anything else. But scientists need to share and compare observations. Therefore, they must have a standard unit of mass that everyone agrees upon. That standard unit of mass is the kilogram. The kilogram is one of the units of *Le Système International d'Unités*. This system is often abbreviated SI. The standard kilogram is housed at the International Bureau of Weights and Measures near Paris, France.

Cross-Curricular Integration

Language arts:
*   Read aloud or suggest that students read one or more of the following books:
    *   *The Three Billy Goats Gruff* (any version)
        Two little goats tell the mean troll that their big brother is coming and would make a bigger meal for him, and they persuade the troll to let them cross his bridge. Big brother goat comes and butts the troll into the river.
    *   *Who Sank the Boat?* by Pamela Allen (Putnam, ISBN 0698-20679-7)
        Good animal friends go for a boat ride, but the boat sinks. Someone sank the boat!

Math:
*   Have students practice measuring and estimating.
*   Have the students make a bar graph of results.

References

David, A. *Understanding Mathematics Five Teachers' Resource Book;* Shortland: New Zealand, 1984; pp 107–109.

Ferruggia, A. et al. *Silver Burdett Science;* Teacher Ed.; Silver-Burdett: Morristown, NJ, 1987; pp 66–67.

Hewitt, P.G. *Conceptual Physics,* 5th ed.; Little, Brown: Boston, 1985, pp 125–142.

Contributors

Sally Drabenstott (activity developer), Sacred Heart, Fairfield, OH; Teaching Science with TOYS peer writer.

Judy Piehowicz, Fairfield Elementary School, Pickerington, OH; Teaching Science with TOYS, 1993–94.

Handout Master

A master for the following handout is provided:

• Measuring Mass in Bears—Data Sheet

Copy as needed for classroom use.

MEASURING MASS USING A PAN BALANCE

Measuring Mass in Bears—Data Sheet

How many counting bears do you need to balance each object?

Material	Mass in Counting Bears
sand	
rice	
puffed rice	
unpopped corn	
popped corn	
small buttons	

RAMPS AND CARS

Ready-set-go! Students explore the effect of different angles of inclined planes on speed and distance.

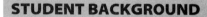

A car racing down a ramp

GRADE LEVELS

Science activity appropriate for grades K–3
Cross-Curricular Integration intended for grades K–3

KEY SCIENCE TOPICS

- distance
- gravity
- inclined planes
- motion
- speed

STUDENT BACKGROUND

Students should have a basic understanding of gravity and falling objects.

KEY PROCESS SKILLS

- predicting — Students predict the distance that a toy car will travel when run down an inclined track.

- controlling variables — Students vary the potential energy of the car by varying the incline height of the track.

TIME REQUIRED

Setup	10 minutes
Performance	30 minutes
Cleanup	5 minutes

Materials

For the "Procedure"

See "Getting Ready" for details on needed materials.

Per class
- 3 same-sized blocks or books
- Hot Wheels® or Matchbox®-type race car
- any or all of the following items to measure and record distances:
  - meterstick, ruler, measuring wheel, or tape measure
  - a pre-marked measuring strip taped to the floor

- ◦ 4 each of 3 different-colored stickers
- ◦ about 300 feet of adding machine tape
- ◦ crayons or markers
- • a track made from 1 of the following:
  - ◦ at least 3 feet of straight plastic race car track
  - ◦ 3-foot x 4-inch length of corrugated cardboard with the corrugations running the long way
  - ◦ 2 3-foot lengths of 1-inch-wide vinyl threshold strip and tape (1-inch-wide masking tape or something similar)
  - ◦ any other item with a smooth surface and an edge to keep the car from falling off the track
- • (optional) a ramp surface at least as long and wide as the track and thick enough not to bow when propped
- • (optional) a wall chart made in 1 of the following ways:
  - ◦ paper or posterboard mounted on the wall or bulletin board
  - ◦ drawn on the chalkboard

For "Variations and Extensions"

❶ All materials listed for the "Procedure" plus the following:
Per class
- • balls
- • more floor space

❷ All materials listed for the "Procedure" plus the following:
Per class
- • different types of balls (such as Ping-Pong™ ball, Nerf® ball, rubber ball, and tennis ball)

❸ Per group or class
- • wide board
- • toy trucks
- • objects (such as marbles or pennies) to load in trucks

❹ All materials listed for Extension 3

❺ Per group or class
- • materials to construct 2 inclined planes
- • 2 toy trucks of different size and shape

❻ Per group or class
- • materials to construct 2 inclined planes
- • 1 or more wind-up or pull-back cars

❼ Per group or class
- • at least 2 toy cars
- • materials (such as car wax, baby oil, and water) to apply to inclined plane

Safety and Disposal

No special safety or disposal procedures are required.

Make sure each group will have enough floor space to allow cars to come to rest on their own (10 feet to 20 feet beyond the end of the ramp). Usually a classroom or hall has adequate space.

This activity requires three basic components: a track that can be elevated on one end, a means of measuring the distance traveled by the car, and a means of recording and displaying the results of the experiments. These components can be set up in a variety of ways to suit your classroom. Below are some suggestions:

A variety of tracks will work in this activity:

- Darda®, Hot Wheels®, and Majorette® make racetrack sets suitable for this activity; but since only the straight pieces are needed for the activity, buying complex sets is not cost effective. Darda® and Hot Wheels® sell accessory sets with just straight track pieces, but not all toy stores carry them.

- Students may have track at home that they can bring in.

- You can make a racetrack from a 3-foot by 4-inch strip of corrugated cardboard with the corrugations running the long way: Fold the cardboard up about ½ inch from each outside edge to form ridges that will keep the car from falling off the track.

- You can also make a track from vinyl threshold strip available at hardware stores. See Figure 1.

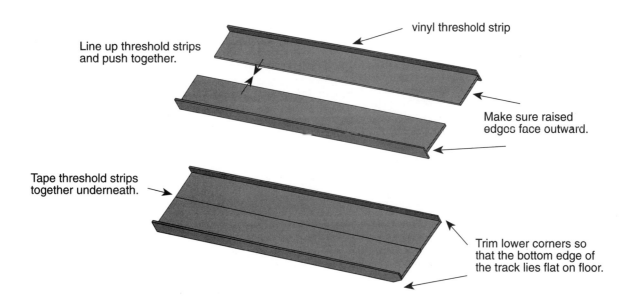

Figure 1: Make tracks using vinyl threshold strips.

If your track bows when propped up, place it on a firmer surface. Anything can be used as a support for the track, but the car must be able to make a smooth transition from the track surface to the floor. We have successfully used two layers of ¼-inch foam core board or corrugated board glued together and angled on one end.

Choose a method for marking and measuring the predicted and trial distances for the three different elevations of the track. Below are some ideas:

- Have students predict how far the car will go and mark that distance on the floor with a colored sticker. Mark the sticker to indicate that this is the predicted distance. After each trial, have them mark the car's position on the floor with the same color of sticker. Repeat, using a different-colored sticker for each elevation of the ramp. Then students can measure the distances using a meterstick, ruler, measuring wheel, or tape measure. Students who cannot yet measure distance with a meterstick can measure distance by counting the numbers of floor tiles traveled by the car.

- Tape a long strip of paper to the floor along the path the cars will run. Measure and mark distance intervals, such as half-meters, along the paper. Have students predict how far the car will go and mark that distance on the paper strip. After each trial, students can use the paper strip to measure the distance the car traveled to the nearest half-meter (or whatever unit you used) and compare to the predicted distance. If desired, students can mark the positions of different cars on the paper strip using crayons, markers, or colored stickers.

- Have students predict how far the car will go and mark that distance on the floor. After each trial, roll a piece of adding machine tape from the end of the ramp to the position of the car. Tape the paper strips in the appropriate position on a chart as shown in Figure 2. If desired, you can have students measure the lengths of the strips. But even without measuring, the relative distances are clearly visible.

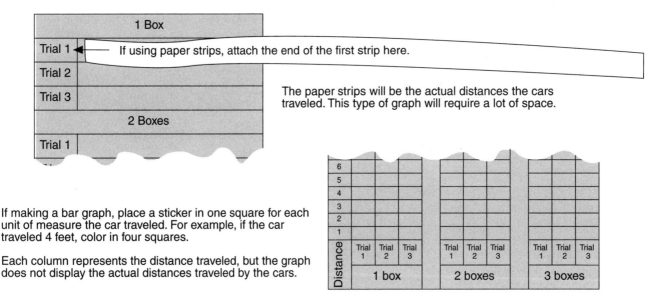

Figure 2: Wall charts can be used to display results in several ways.

Choose a method for recording the experimental data. Some ideas follow:

- If using the adding machine tape idea described above, you will automatically create a graphical display of the results.

- Have students record measured distances and make a bar graph.

- Use a chart like the lower one in Figure 2 to create a bar graph from the measurements: Mark off enough boxes in each column to accommodate your longest measurement. For example, if your students measured by half-meters and your longest distance was 10.5 m, you would need at least 25 boxes. Instruct students to color or place a sticker in one box for each half-meter the car traveled.

Introducing the Activity

Students must be told that while science activities are fun, this is not a play activity. They must follow directions and move the cars only when instructed to do so.

Procedure

1. Place one block under one end of the track or board. (If using a board, place the racetrack on the board and adjust the track so it extends beyond the edge of the board to touch the floor.)

2. Ask students to predict where the car will stop after being released from the top of the ramp.

3. Mark, measure, and record the predicted distance as desired.

4. Have a student place a car on the track at the top of the ramp and then release it. Make sure the student does not push the car.

5. After the car comes to rest, have students mark, measure, and record the distance as desired.

6. Have students release the car two more times to see if it stops at about the same place, marking, measuring, and recording the distance as desired.

7. Add one more block under the ramp for a total of two blocks, and repeat Steps 1–6.

8. Add one more block under the ramp for a total of three blocks, and repeat Steps 1–6.

Class Discussion

1. Have students compare their predicted distances with the actual distances.

2. Have students consider the distances traveled by cars starting at different heights. Ask students to draw a conclusion about the relationship between the height of the ramp and the distance traveled by the car.

3. As a class, discuss the following questions:

   a. Did any of the cars go farther than the predictions?
   b. Did any of the cars not go as far as the predictions?
   c. Which car traveled the farthest?
   d. Which car stopped closest to the ramp?
   e. Why do you think one car went farther than another?

Variations and Extensions

1. Balls have less friction than cars and could be used instead of cars. More floor space will be needed for the balls to come to rest.

2. Instead of changing the angle of the incline, keep the angle of the incline constant, and roll balls with different masses down the incline.

3. Use a wide board to make an inclined plane. Have races with identical toy trucks carrying different amounts of mass in their cargo areas. Predict the winners.

4. Put a road block in front of the moving trucks described in Extension 3. What happens to the trucks and to the contents?

5. Construct two inclined planes that touch the floor opposite each other, about 1 m apart. Release a toy truck from each incline and observe which one reaches the bottom first and which one bounces the farthest when they collide.

6. Set up two inclined planes to represent an open drawbridge. Use one or more wind-up or pull-back cars to see if it is possible to jump the gap and land on the other side. (See Figure 3.) How far can the car(s) jump?

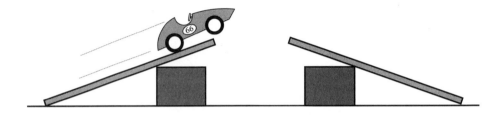

Figure 3: Try to jump a car over a gap between two inclined planes.

7. Have races using toy cars and inclined tracks. How can the cars be made to go faster? How can the surface be prepared to increase the speed of the cars? (See "Materials" for suggested materials to apply to the track.)

Explanation

 The following explanation is intended for the teacher's information. Modify the explanation for students as required.

In this activity, you observe that the toy car rolls farther as the incline of the track becomes steeper. The car rolls farther with a steeper incline because it is moving faster when it reaches the bottom of the ramp. The faster the car is moving when it leaves the ramp, the more distance it can cover before friction slows and stops it.

So why is the car moving faster when it reaches the bottom of a steeper incline? To understand why, we need to think about how the force of gravity acts on objects. If you hold a toy car in the air and drop it, it falls to the ground. When you place the same toy car at the top of an inclined ramp and let go, the car rolls down the ramp. In both cases the car moves because the force of gravity acts on it. This force of gravity is the weight of the car. But the car falls to the ground much faster than it rolls down the ramp. Why? In addition to a small effect from friction, there is another important factor.

When you hold the car in the air and drop it, the entire force of gravity (or weight) acts in the same direction—straight down—causing the car to accelerate and reach the ground quickly. (See Example 1 in Figure 4.) But in this experiment, the downward force of gravity is divided into components that act in more than one direction. When the car is on the ramp, a part of the force of gravity acts parallel to the ramp, causing the car to accelerate. But this acceleration is not as great as when the car falls straight down, because the force causing the acceleration is not as great. In Figure 4, the arrows labeled "B" represent the forces causing acceleration. (See the arrows labeled "B" in Figure 4. Note the difference in size.) The rest of the force of gravity acts at right angles to the ramp and holds the car against the surface. (See the arrows labeled "A" in Figure 4.)

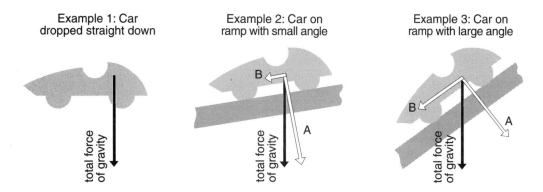

Figure 4: When the car is dropped, all components of the force of gravity act downward. On an incline, component A of the force of gravity holds the car against the surface of the ramp, and component B acts along the plane of the ramp to accelerate the car.

As the angle of the ramp increases, the component of the force of gravity that acts parallel to the ramp increases, and the car's speed increases more quickly as it rolls down the ramp. When the car reaches the end of the ramp it is going faster than it does when the ramp angle is smaller. Even though the component of the force of gravity that acts parallel to the ramp increases as the ramp angle increases, the total force of gravity on the car does not change. Thus, the component of the force of gravity that holds the car against the surface of the ramp must decrease as the ramp angle increases. (See Figure 4.)

You can illustrate this phenomenon in a simple demonstration. Begin by attaching a rubber band to the back of a car. Place this toy on the incline and while holding the rubber band in one hand raise the incline with the other. Notice that the rubber band stretches more and more as you raise the board. (See Figure 5.) This indicates that as the angle of the incline increases, more of the weight of the car points down the incline.

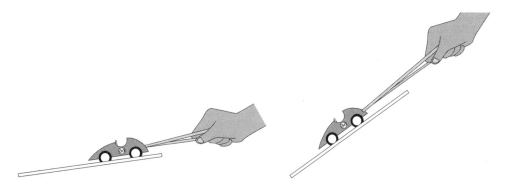

Figure 5: As the slope increases, so does the pull on the rubber band.

Cross-Curricular Integration

Language arts:
- Read aloud or suggest that students read one or more of the following books:
  - *The Lazy Bear,* by Brian Wildsmith (Oxford University, ISBN 0-19-272158-5)
    A bear learns the effects of an increased angle of incline when he rides a cart down the steep hill.
  - *The Mouse and the Motorcycle,* by Beverly Cleary (Morrow, ISBN 0-688-31698-0).
    A mouse "borrows" a boy's toy motorcycle and causes some surprises in the hotel where he lives.
  - *The Way Things Work,* by David Macaulay (Houghton Mifflin, ISBN 0395428572)
    This reference book for older children explores the mechanics of movement with wonderful diagrams and explanations.

Math:
- Have older students construct bar graphs of their results.

References

Activities Handbook for Teachers of Young Children, 5th ed.; Houghton-Mifflin: Boston, 1990.

Butzow, C. *Science through Children's Literature;* Teacher Idea: Englewood, CA, 1989.

Contributors

Sally Drabenstott (activity developer), Sacred Heart, Fairfield, OH; Teaching Science with TOYS peer writer.

Jill Stelzer, Murlin Heights Elementary School, Dayton, OH; Teaching Science with TOYS, 1993–94.

BALLOON ON A STRING

What causes the balloon to travel along a path without a person pushing or pulling it?
In this activity, students discover that an object can be pushed or pulled by an invisible force.

A balloon on a string

GRADE LEVELS

Science activity appropriate for grades K–3
Cross-Curricular Integration intended for grades K–3

KEY SCIENCE TOPICS

- force
- motion

STUDENT BACKGROUND

Students should have completed some previous activities that let them experience that a force is a push or pull. For example, students could blow on a pencil to see that moving air exerts a force to move the pencil across the desk.

KEY PROCESS SKILLS

• observing	Students observe the behavior of an inflated balloon as the air escapes rapidly from the opening.
• predicting	Students use their observations to predict behavior in a similar case where the balloon is modified to move freely along a string guide.

TIME REQUIRED

Setup	15	minutes
Performance	20–30	minutes
Cleanup	5	minutes

Materials

For the "Procedure"
Per class
- 3 or 4 large, long, narrow balloons

These balloons, sometimes packaged as "airship" balloons, should be approximately ½ inch x 6 inches uninflated, and at least 4 inches x 12 inches inflated.

- ball of string

- 1 plastic straw
- tape
- materials to attach string to wall, blackboard, or bulletin board such as tape, a hook, a magnetized clip, or a thumbtack
- (optional) a rope with which to play tug-of-war
- (optional) a pair of roller skates or roller blades

For "Extensions"

❷ All materials listed for the "Procedure" plus the following:
Per class
- different-shaped balloons
- different kinds of string such as yarn, fishing line, and twine
- materials to attach several strings to wall, blackboard, or bulletin board

❹ All materials listed for the "Procedure" plus the following:
Per group
- ruler or meterstick

Safety and Disposal

No special safety or disposal procedures are required.

Getting Ready

Attach a 20-foot (6-meter) length of string to a wall, blackboard, or bulletin board at one end of the classroom.

Introducing the Activity

Options:

- A game of tug-of-war in the gym or outside could be used to illustrate forces.

- Have someone bring in roller skates or roller blades. While wearing the roller skates, have that person throw a heavy object. He or she will roll the other way.

Procedure

1. Blow up a balloon (but don't tie it shut) and ask students, "What is the difference in the appearance of the balloon before and after it was blown up?" *The balloon is bigger.* Explain that balloon material is elastic—it stretches when air is forced into the balloon.

2. Tell students, "Predict what will happen if I let go of the balloon." After hearing students' answers, let go of the balloon and have students observe what happens.

3. Ask students, "Why is the air pushed out of the balloon?" After hearing students' answers, summarize by saying, "The stretched material of the balloon exerts a force on the air inside, causing the air to be pushed out of the hole in the balloon."

4. Ask students, "Why does the balloon move forward?" After hearing students' answers, summarize by saying, "The air expelled backward exerts an equal force on the balloon, pushing it forward."

5. State Newton's second law of motion: "If a force is exerted on a stationary object that is free to move, that object will move in the direction of the force." (Pushing a toy car is an example.)

6. State Newton's third law of motion: "The balloon exerts a force on the air, making the air move. The air exerts a force on the balloon, making the balloon move."

7. One end of the string is already attached to the wall. Feed the other end of the string through the plastic straw. Attach the loose end of the string to the wall on the other side of the room.

8. Put two pieces of tape on the straw as shown in Figure 1.

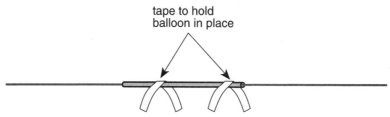

Figure 1: Place two strips of tape on the straw.

9. Blow up a balloon and hold it closed. Fasten the balloon to the tape. (See Figure 2 for completed assembly.) Ask students, "What will happen when I let go of the balloon this time?" *The balloon will move forward.*

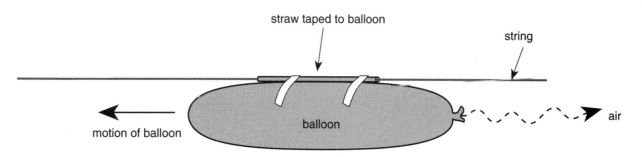

Figure 2: The balloon moves along the string.

10. After making sure that the string is stretched as tightly as possible, let go of the balloon to check the students' prediction. Ask students, "How does this balloon move differently than the first balloon we let go?" *The first balloon went every-which-way, but this balloon moves in a straight line along the string. The string acts like a track, guiding the balloon in a straight line.*

Extensions

1. Vary the amount of air in the balloon to see what effect this has on the distance traveled by the balloon.

2. Set up several lines across the room and have balloon races:

   a. Try different-shaped balloons.

   b. Try different kinds of string.

3. See if the balloon will move up an inclined string.

4. Have students measure the distance the balloons travel using a ruler or meterstick and construct a graph showing the data gathered from several trials.

Explanation

 The following explanation is intended for the teacher's information. Modify the explanation for students as required.

These demonstrations illustrate the principle of simple rocket propulsion. Two of Newton's laws apply here. Newton's second law states that if a force is exerted on a stationary object that is free to move, then the object will move in the direction of the force. Young children should not have difficulty with this concept when it is explained along with demonstrations.

Newton's third law states that when one object, A, exerts a force on another object, B, object B exerts an equal and opposite force on object A. This statement of Newton's third law is a generalization that young children would not be able to grasp in the abstract. They can understand the concept with a specific demonstration and the statement in Step 6 of the "Procedure."

Assessment

Have students complete the "What Made It Go?" Assessment Sheet (provided).

Cross-Curricular Integration

Language arts:
- Read aloud or suggest that students read one or more of the following books:
  - *The Blue Balloon,* by Mick Inkpen (Little, Brown, and Co., ISBN 0-316-41886-2)
    A little boy finds a blue balloon that becomes very special to him. This book has fold-out pages to make the balloon grow.
  - *I'm Flying,* by Alan Wade (Alfred Knopf, ISBN 0-394-84510-2)
    A little boy floats away on his balloon across mountains, plains, cities, and the sea until he lands on a desert island.
  - *Mirandy and Brother Wind,* by Patricia McKissack (Alfred A. Knopf, ISBN 0-0394-98765-9)
    To win first prize in the Junior Cakewalk, Mirandy tries to capture the wind for her partner.
  - *The Red Balloon,* by Albert Lamorisse (Doubleday, ISBN 0-385-00-343-9)
    A little French boy follows his red balloon as it travels over Paris, France.
- Try a creative writing activity—"If I could ride on a balloon…" Have students write their stories on paper cut-outs of balloons.

Reference

Faughn, J.S.; Turk, J.; Turk, A. *Physical Science;* Saunders College: Philadelphia, 1991; pp 36–44.

Contributors

Sally Drabenstott, Sacred Heart, Fairfield, OH; Teaching Science with TOYS peer writer.

Cathy Farmer, Sabina Elementary School, Sabina, OH; Teaching Science with TOYS, 1993–94.

Pat King, Fairfield Central School, Fairfield, OH; Teaching Science with TOYS, 1992.

Gerri Westwood, Berkeley Elementary School, Spotsylvania, VA; Teaching Science with TOYS, 1992.

Handout Master

A master for the following handout is provided:

- What Made It Go?—Assessment Sheet

Copy as needed for classroom use.

BALLOON ON A STRING

What Made It Go?—Assessment Sheet

1. In the space below, draw what you saw when we did the "Balloon on a String" activity.

2. You just drew a picture of our "Balloon on a String" activity. Describe what you saw.

   _____

   _____

   _____

3. Why do you think this happened?

   _____

   _____

   _____

4. What if you used a balloon in a different size? Predict what would happen.

   _____

   _____

   _____

5. What other things might act the same way?

   _____

   _____

   _____

PING-PONG™ PUFFER

Students observe the effect of air pressure while "puffing" Ping-Pong™ balls.

A student "puffs" a Ping-Pong™ ball

KEY SCIENCE TOPICS

- air
- air pressure
- gravity

KEY PROCESS SKILL

- inferring

Students infer that moving air has a lower air pressure than air that is not moving.

TIME REQUIRED

Setup	20	minutes
Performance	30	minutes
Cleanup	5	minutes

Materials

For the "Procedure"

Per student
- plastic cap from a soft-drink bottle
- 1 bendable straw with flex-joint
- 1 Ping-Pong™ ball

Per class
- 1 drill with bit the same diameter as the straws
- 1 hot melt glue gun with glue
- 1 commercial Ping-Pong™ Puffer

Commercial Ping-Pong™ Puffers are known by many different names. A Blow Pipe is available from U.S. Toy Company, (800) 255-6124; a Blow Cup with Ball is available from Oriental Trading Company, (800) 228-2269.

For "Variations and Extensions"

 Per class
- beach ball
- fan

❸ Per class
  • beach ball
  • shop vacuum

Safety and Disposal

Be certain that students take short rests between tries due to the possibility of hyperventilation. No special disposal procedures are required.

Getting Ready

1. In the center of each plastic cap, drill a hole of the same diameter as the straws.

2. Make one sample of the Ping-Pong™ Puffer by inserting the short end of the bent straw through the cap. (See Figure 1.) Make sure the straw is pushed up far enough through the cap so that it will contact the ball when the ball is placed on the open end of the cap.

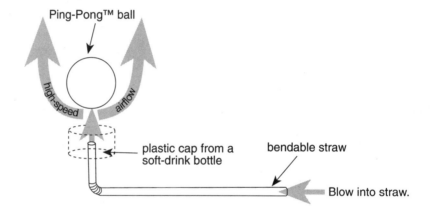

Figure 1: Build a Ping-Pong™ Puffer.

3. Practice using the puffer as follows: Hold the end with the cap on it. Put the long end toward your mouth. Hold the puffer level while you place the ball on the cap and balance it there. Take a big breath, and blow strongly and steadily through the straw.

If the ball spins but does not rise, slide the cap down the straw slightly and try again. If the ball just falls off when you blow, the straw may be too high in the cap or you may not be holding the puffer level enough.

4. When you have the cap properly positioned, secure the cap to the straw with the glue gun.

Introducing the Activity

Before doing this activity the students should understand that air is "something" even though we cannot see it. To help them understand this idea, students should do the following: trap air in a bag and feel the resistance as they squeeze it; make parachutes to catch air; trap air in an inverted cup as it is lowered into water; and feel the force of a light breeze push against a plastic sheet as they hold it. After students are positive that air is something that can exert force and be moved (instead of being an invisible nothing), they are ready for the Ping-Pong™ Puffer.

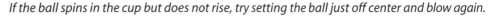

Part A: Making and Using the Ping-Pong™ Puffer

1. Show students the Ping-Pong™ Puffer assembled in "Getting Ready" and a commercial Ping-Pong™ Puffer.

2. Ask students to predict what will happen if someone blows hard through the straw while the Ping-Pong™ ball is balanced on the cap.

3. Demonstrate, with the ball rising and coming back to rest on the cap. Have students share observations.

If the ball spins in the cup but does not rise, try setting the ball just off center and blow again.

4. Have students construct their own puffers and try them.

5. Have them experiment with moving the cap slightly higher or lower on the straw and observe any changes.

6. When students have the caps properly positioned, secure the caps to the straws with the glue gun. Allow students more time to practice with the Puffers.

Part B: Turning the Ping-Pong™ Puffer Upside-down

1. Have students try holding the puffer with ball upside-down when they are not blowing into it. Ask, "What happens?" *The ball falls out.*

2. Show students that the puffer can still work while turned upside-down: Turn the puffer upside-down, hold the ball near the cap, and start to blow. Then let go of the ball. It should stay suspended in the cup. Allow students time to practice this with the puffers.

3. Ask, "Why doesn't the ball fall out?" Through a facilitated discussion, lead students to suggest that something might be holding the ball in place. Tell students that they will be trying a couple of experiments to help them figure out what is keeping the ball from falling.

4. Have students hold a strip of paper in front of their mouths as shown in Figure 2. Ask, "What does the paper do?" *The paper droops down.* Now have students blow across the paper. "What happens now?" *The paper rises.* "Where is the air moving faster, above the paper or below?" *Above.*

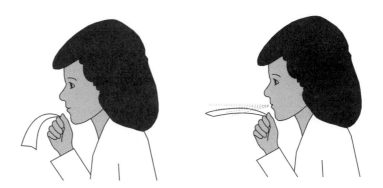

Figure 2: When you blow across the top surface, the paper rises.

5. If you have access to a sink, demonstrate the following: Tape a Ping-Pong™ ball to a string and allow the ball to swing into a stream of running water. Tug gently to one side. The ball will remain in the stream. (See Figure 3.) Ask, "What is moving faster, the water or the air near the water?" *The water.*

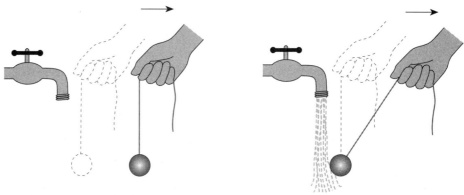

When no water is flowing, the ball moves away from the faucet freely.

When water is flowing, the ball remains in the stream, even if you tug gently.

Figure 3: The higher pressure of the stationary air next to the ball pushes it into the region of reduced pressure near the water stream.

6. Have students try their puffers upside-down again. Ask, "Where is the air moving faster?" *Around the areas of the ball closest to the cap.* (See Figure 4.)

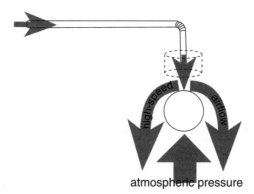

Figure 4: Atmospheric pressure keeps the ball from falling.

7. Through a facilitated discussion, lead students to observe that in each case, the fluid (either air or water) in one area was moving more quickly that the fluid in an adjacent area. Continue to discuss as much more content as appropriate for your students. (See the "Explanation.")

Variations and Extensions

1. The same activity can be done without the cap on the bent straw. In this case, the ball must be placed above the straw after the airstream has begun moving.

2. Try suspending a beach ball in the airstream of a fan: Tilt a circular fan so it blows upward. Place a beach ball on the stream of air to see if you can suspend it there. This will work with some fans and balls but not all.

3. Try suspending a beach ball in the airstream of a shop vacuum: Insert the hose into the exhaust port and point it upward. Find a beach ball of a size and weight that can be suspended on the airstream.

Explanation

The following explanation is intended for the teacher's information. Modify the explanation for students as required.

When you place the Ping-Pong™ ball in the cup of the puffer, the force of gravity pulls down on it and keeps it sitting in the cup. When you blow in the puffer, the air flowing out of the straw pushes the Ping-Pong™ ball upward with sufficient force to overcome the force of gravity. (See Figure 1.) When you first begin to blow, the air causes the ball to be lifted and quiver until it is centered on the stream of air coming through the straw. The steady stream of air evenly distributed around the ball results in a force that keeps the ball centered just above the cup. The ball stays securely in the airstream even if the direction of the airstream changes a bit. The beach ball in Extensions 2 and 3 can be suspended on the airstream of a fan or shop vacuum for the same reason.

When you turn the puffer upside-down, some new factors become important. As you saw, the ball falls out of the upside-down puffer when no air is being blown through it. It falls because the force of gravity pulls it to the ground. If the ball is to stay in the upside-down cup, some other force must be oppose the force of gravity. What is it? The paper strip and faucet activities in Part B of the "Procedure" provide some clues. In each case, the fluid (either air or water) in one area is moving more quickly than the fluid in an adjacent area. The paper strip and the ball on a string move toward the area of faster-moving fluid. Why? Because as the speed of a fluid increases, the pressure in the fluid decreases, creating an area of low pressure. Then, the relatively higher pressure in surrounding areas pushes the object towards the low-pressure area. This is clearly visible as the paper moves up or the ball on a string moves towards the water stream.

The effect of pressure differences also explains why the Ping-Pong™ ball does not fall out of the cup when the puffer is turned upside-down while blowing. The faster-moving air around the areas of the ball closest to the cap creates an area of low pressure. The relatively higher atmospheric pressure exerted against the area of the ball farthest from the cap is enough to oppose the force of gravity.

Cross-Curricular Integration

Language arts:
- Have your students pretend that their Ping-Pong™ Puffers are really magical ships with launch pads. In their ships, students can travel anyplace they want to go. Have the students write stories about one of their trips and draw illustrations for their stories.

Contributors

Mark Beck (activity developer), Indian Meadows Primary, Ft. Wayne, IN; Teaching Science with TOYS peer mentor.

Mary Beth Field, Northern Elementary School, Butler, KY; Teaching Science with TOYS, 1993–94.

BALANCING STICK

*Students build and explore balance toys,
and discover how varying the amount and position of mass affects the toy's balance.*

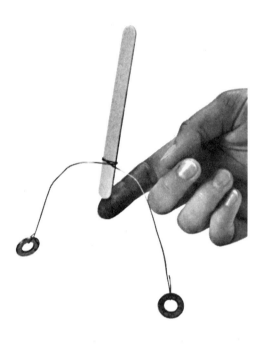

A Balancing Stick

GRADE LEVELS

Science activity appropriate for grades K–6
Cross-Curricular Integration intended for grades K–3

KEY SCIENCE TOPICS

- balance
- center of gravity

STUDENT BACKGROUND

With younger students (K–1), this activity is most effective if they have had some previous experience balancing items such as a small board on a fulcrum or a ruler across a finger.

KEY PROCESS SKILLS

- hypothesizing   Students form and test hypotheses of how to balance craft sticks on end.

- investigating   Students investigate how changes in the positions of parts of the Balancing Stick affect balance.

TIME REQUIRED

Setup	20	minutes
Performance	20	minutes
Cleanup	5	minutes

Materials

For "Getting Ready" only
The wire cutters are intended for teacher use only.

- wire cutters

For the "Procedure"
Per student
- 1 Popsicle™ or craft stick
- 1 12-inch length of 20- or 22-gauge soft wire

- 1 of the following to use as weights:
  - 2 paper clips
  - 2 hex nuts
  - 2 ⅜-inch flat washers
- (optional) additional weights to use as added mass in Part B, Step 2, of the "Procedure"
- 1 pencil with an eraser
- 2 pieces of masking tape, each about 2 inches long
- 2 paper clips

For "Variations"

❶ Per student
- 1 Clown-on-a-Ball (master provided)
- rubber cement or glue
- crayons
- scissors

❷ Per student
- 1 Headstanding Clown (master provided)
- rubber cement or glue
- crayons
- scissors

Safety and Disposal

Students should take care to work in their own space when bending wire, keeping it well away from their eyes and the eyes of other students. You may wish to bend a ½-inch hook on each end of each student's wire ahead of time to avoid this concern. No special disposal procedures are required.

Getting Ready

1. Cut the wire with wire cutters.

2. (optional) Bend each wire ½ inch from each end to form two hooks.

3. Put a pencil mark on each Popsicle™ stick about 1¼ inch from one end. (This could be done by students who have learned to measure length.)

4. Prepare one Balancing Stick according to the "Procedure."

Introducing the Activity

Distribute a Popsicle™ stick to each student. Tell the students you have been trying to balance the stick on your fingertip and that you would like them to show you how they would do it. Most will probably place the flat edge of the stick across a finger at the midpoint of the stick. Acknowledge their skills, then tell them you actually wanted to balance it on one end. Let them try for a little while. Then show them the Balancing Stick prepared in "Getting Ready." Balance it on a finger. Ask the class to describe what they observe. Let them speculate about why the regular Popsicle™ stick would not balance on end but did balance when the wire and washers were added.

Tell the class that they will test their hypotheses about how the stick balances by building their own Balancing Sticks (in Part A of the "Procedure") and trying the three "Balancing Challenges" (in Part B of the "Procedure").

Procedure

Part A: Making the Balancing Stick

Have each student prepare a Balancing Stick by following these steps:

1. Bend the wire approximately in half.

2. Wrap wire around the stick about two times at the pencil mark, ending up with the two ends of the wire about the same length.

3. Curve the wire ends down, away from the long end of the stick.

4. (optional) If you did not bend the wires in "Getting Ready," have students bend the wire about ½ inch from each end to form two small hooks.

5. Add a weight to each hook and bend the hook shut so the weight will not easily fall off. (See Figure 1 for completed assembly.)

6. Try balancing the stick on a fingertip. If it tips over and falls off, observe which way it tipped. Try bending one arm slightly in or out and observe how it balances now. Keep trying.

7. Tape a pencil flat on the table with the eraser end sticking out from the table. Balance your stick on the eraser. If needed, make minor adjustments by slightly bending wire arms.

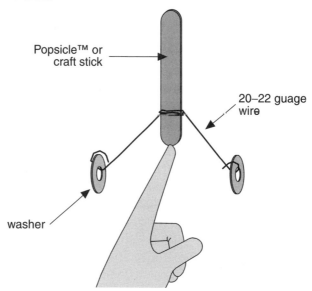

Figure 1: Assemble and balance the stick.

Part B: Balancing Challenges

1. Challenge students to balance their sticks horizontally with the resting point far to one end (like a diving board), supporting the stick at the end, not in the middle. (See Figure 2.) (To do this, they must bend the wire to a different position.)

   Hint: Shift the mass in the direction the stick should tip.

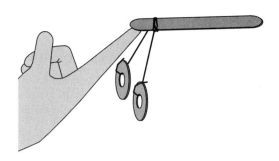

Figure 2: Balance the stick horizontally.

2. Tell the students to put the wire back into its original position so that the stick balances straight up (vertically). Challenge them to balance their sticks horizontally by adding mass to or removing mass from one or both wire arms.

3. Have students balance their sticks vertically again. Challenge them to balance the sticks vertically after adding more mass to the wire hooks. Can they balance the sticks vertically with less mass on the wire hooks? Have them try paper clips instead of washers.

Variations

1. Students can color the front side of the Clown-on-a-Ball and cut it out. Have them use two dots of rubber cement or glue on the flat part of the stick, and attach it to the back side of the cut out. The Popsicle™ stick should extend from near the foot down to the edge of the ball, with the wires bending out of the ball. Now, balance the Clown-on-the-Ball on a pencil eraser or fingertip. If students used rubber cement, they can peel off the clown and use the same stick for Variation 2.

2. Have students prepare the large Headstanding Clown as they prepared the Clown-on-a-Ball in Variation 1. Have them glue the cutout onto the Balancing Stick so that the lower end of the stick is even with the top of the clown's head and the wires follow the clown's arms.

Explanation

> *The following explanation is intended for the teacher's information. Modify the explanation for students as required.*

In "Introducing the Activity," your students try to balance the stick on end with no other mass added. Under these conditions, the stick's center of gravity is well above the resting point at the lower end of the stick, and the stick falls. The Balancing Stick (the stick with wire and washers added) balances because the additional mass moves the center of gravity to a point below the lower end of the stick.

Changing the position of the added mass changes the behavior of the Balancing Stick. For example, if the wire arms are pushed down and nearly together below the lower end of the stick, the center of gravity is lowered even more. The toy still balances but becomes less stable. If one of the wire arms is bent upward and outward, the center of gravity of the whole system is redistributed and the center of gravity moves upward and outward, tilting or rotating the whole system in the opposite direction to move the center of gravity back down. (Try this maneuver to see how it works.) If the center of gravity moves up to a point above the lower end of the stick, the whole system may rotate off its resting place.

Changing the amount of mass also changes the behavior of the Balancing Stick. Adding more mass to the arms in the lower position lowers the center of gravity and makes the toy easier to balance. Removing mass from the lowered arms raises the center of gravity, making the Balancing Stick harder or impossible to balance.

Assessment

Options:

- After completing the activity, observe each student's ability to balance the stick on his or her finger. Then ask some of the following questions: "Did the stick by itself balance on your finger? Why or why not? What did we add to it? Where did we place the extra mass to help balance the stick?"

- Set up the learning center described below, including materials, a statement of the challenge, and a diagram showing the end result (if desired).

Styrofoam™ Ball Balancers

Materials: Styrofoam™ ball, 2 straws, toothpick, enough clay or Play-Doh® to make two small balls

Challenge: Stick a toothpick into a Styrofoam™ ball. Get the ball to balance (toothpick down) on the top of a pencil eraser. Hint—Insert two straws into the ball at different angles and add mass to the ends of the straws.

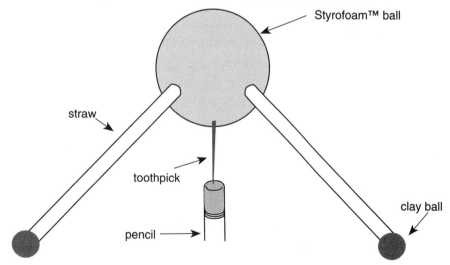

Figure 3: Balance a Styrofoam™ ball on a pencil eraser.

Cross-Curricular Integration

Art:
- Have the students draw their own balanced figures. These figures could be circus performers or other people or animals that are good balancers.

- Have the students make a mobile. (See Figure 4.) They can use hangers, plastic straws, pennies, and fishing line cut in various lengths. Have each student choose a theme for his/her mobile and collect objects or pictures to carry out the theme. Have them attach one end of each piece of fishing line to an object and the other end to the end of a straw. Have them tie an object to the end of each straw. Then have the students hang the straws from the hanger or from each other. When the students are finished, have them tape pennies or washers on the objects to make them balance. (The hanger and the straws should be level.)

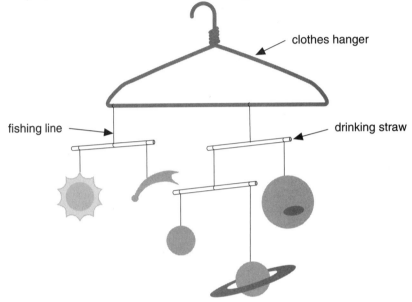

Figure 4: Make a mobile.

Language arts:
- After students have completed this activity, show them the cover of *Mirette on the High Wire,* by Emily Arnold McCully (Putnam, ISBN 0-399-22130-1). Ask, "What is Mirette doing? Why is she holding her arms out?" Read the book to the students. Ask them to look for pages that show balancing. After reading the story, you may wish to have the students design Mirette cutouts for their Balancing Sticks, similar to the clowns used in Variations 1 and 2.
- Have the students write a cinquain poem about their Balancing Sticks. In the lower grades, the cinquain can be written as a whole class activity. By third grade, most students can compose their own. If the students do the first art integration, have them write their poems around the outline of their drawn figures. The standard form for a cinquain is as follows:

Noun
Adjective, adjective
Three-word sentence
Four participles
Noun

A sample poem for this activity would be as follows:

Mass
Upside-down, silly
Will it balance?
Moving, falling, wobbling, standing
Center of gravity

- Read aloud or suggest that students read one or more of the following books:
  - *Bearymore,* by Don Freeman (Puffin, ISBN 0-14-050-279-3)
    A bear learns to ride a unicycle on a tightrope.
  - *The Napping House,* by Audrey Wood (Harcourt Brace Jovanovich, ISBN 0152567119)
    People and animals in a house balance on top of each other like a pyramid.

Life science:

- Have the students study how we can change the center of gravity of our bodies to remain balanced while walking, running, standing, or bending over.
- Discuss the use of crutches or prosthetic limbs to maintain balance and mobility.

Physical education:

- Have the students take turns walking on a balance beam. Then have them try walking on the beam while carrying a heavy weight in one hand.

Social studies:

- Have students study the history of the circus, especially the history of balancing acts.

Reference

Hewitt, P. *Conceptual Physics;* Addison-Wesley: Menlo Park, CA, 1987; pp 119–133, 191–194.

Contributors

Mark Beck (activity developer), Indian Meadows Primary School, Ft. Wayne, IN; Teaching Science with TOYS peer mentor.

Mary Buck, Claymont Elementary School, Ballwin, MO; Teaching Science with TOYS, 1992.

Patricia Carlin, Pierremont Elementary, Manchester, MO; Teaching Science with TOYS, 1992.

Tammy Cihaski, Hawthorn Hills School, Wausau, WI; Teaching Science with TOYS, 1993.

Cherie Kuhn, Tri-County North Elementary, Lewisburg, OH; Teaching Science with TOYS, 1992.

Jill Rowe, Holy Family School, New Albany, IN; Teaching Science with TOYS, 1992.

Paul Seiser, Hawthorn Hills School, Wausau, WI; Teaching Science with TOYS, 1993.

Carol Shay, Valley View Elementary, Austin, TX; Teaching Science with TOYS, 1993.

Susan Thomas, Sixth Grade Center, Cedar Hill, TX; Teaching Science with TOYS, 1993.

Handout Master

A master for the following handout is provided:
- Clown Templates

Copy as needed for classroom use.

BALANCING STICK
Clown Templates

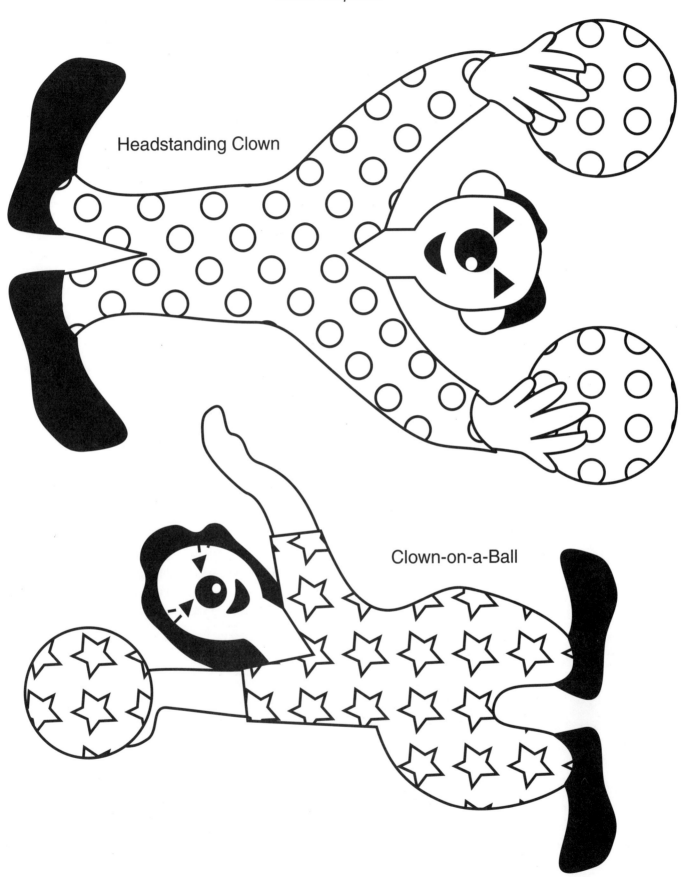

Headstanding Clown

Clown-on-a-Ball

THE SKYHOOK

Students explore gravity and balance while playing with the Skyhook, a popular pioneer toy.

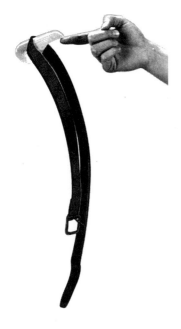

A Skyhook

GRADE LEVELS

Science activity appropriate for grades K–6
Cross-Curricular Integration intended for grades K–3

KEY SCIENCE TOPICS

- center of gravity
- gravity

STUDENT BACKGROUND

Students can work on this project without prior experience; however, some experience with balance toys may add to their understanding of the concept.

KEY PROCESS SKILLS

• observing	Students observe the behavior of the Skyhook.
• hypothesizing	Students form and test hypotheses about how the Skyhook works.

TIME REQUIRED

Setup	5	minutes (plus 5 minutes per hook if you are cutting out the hooks)
Performance	30	minutes (plus 5 minutes if students sand the hooks)
Cleanup	5	minutes

Materials

For "Getting Ready" only

The materials below, intended for teacher use only, are needed to make Skyhooks the first time the activity is done. Alternatively, Skyhooks can be ordered from Frank Guzik, (219) 432-1519.

- 1 jigsaw or saber saw with a scrolling blade
- 1 7-inch x 3-inch piece of oaktag or paperboard

This can be cut from a cereal box.

- length of 1-inch x 2-inch pine or similar easy-to-work wood

A 6-inch length of wood makes 2 hooks. Each group of students will need 1 hook.

For the "Procedure"

Per group of 4 students
- 1 leather belt

If the leather belt is too rigid, the Skyhook will not work.

- (optional) 1 piece of sandpaper about 3 inches square
- 1 Skyhook

If desired, a wooden clothespin without a spring can be used instead of a Skyhook.

For "Extensions"

❶ All materials listed for the "Procedure" plus the following:
- cloth belt
- 1 leather belt with large heavy buckle

❷ All materials listed for the "Procedure" plus the following:
- 1 6-inch length of string

Safety and Disposal

Adults or upper-level shop class students should cut out the hooks and be sure to follow any safety precautions pertinent to the tools being used. No special disposal procedures are required.

Getting Ready

Transfer the Skyhook pattern (See Figure 1) to the oaktag or paperboard. Trace a Skyhook for each group onto the length of wood. Cut out the Skyhooks. Practice balancing the Skyhook and belt on your finger as shown in Figure 2.

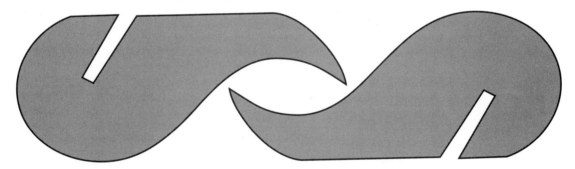

Figure 1: Transfer the Skyhook patterns to oaktag or paperboard.

Introducing the Activity

Show the Skyhook to the class. Explain that it is an early American toy that pioneer boys and girls played with because it was easy to make and almost magical to use. Explain that the pointed end of the Skyhook can balance on your finger. Now, pretend to try to balance the Skyhook, without anything hanging in the slot. The hook will promptly fall off your finger. Then put a leather belt in the slot and demonstrate again as shown in the photo at the beginning of this activity. Now the Skyhook will balance. Do not explain the activity at this time.

Procedure

Students can record their observations on the "Balancing Belts" Observation Sheet (provided).

1. Have students sand the rough spots off the Skyhooks if needed.

2. Each group should try to balance their hook with nothing in the slot, reinforcing the observation made in "Introducing the Activity" that the hook won't balance on its own.

3. Have each group try balancing the Skyhook with a leather belt in the slot.

4. The students should observe those hooks that are balancing and those that are not and discuss in their groups what conditions need to be met for the hooks to balance.

5. Have students report their observations to the class and offer explanations.

6. Direct students' attention toward the belt, its stiffness, and how it hangs on a finger (flat, not in a slanted position) versus the way it hangs on the Skyhook. Then note where the ends of the belt are positioned when it is on the hook.

7. Continue the discussion. Offer the explanation if necessary.

Extensions

1. Have students try the Skyhook with a soft cloth belt (less mass and less rigid) or a leather belt with a large, heavy buckle (more mass).

2. Have the students in each group hang the belt by a string through the slot.

3. Have students experiment to answer these questions: If you hang the Skyhook with a belt on it on some point in your house where it will not be disturbed, how long will it hang there? Will the hook get tired? Why or why not?

Explanation

The following explanation is intended for the teacher's information. Modify the explanation for students as required.

The Skyhook seems to defy gravity. How does it work? To answer, we must look at the Skyhook and belt as a system and consider the center of gravity of the system. When the Skyhook without a belt is placed on the finger, the center of gravity is beyond the support point on the fingertip. Thus, the Skyhook falls. If you observe carefully after the belt is added, you can see that much of the belt is directly under the support point or even slightly behind it. This moves the center of gravity to a position at or somewhat behind and below the support point, a condition necessary for the hook and belt to balance. So even though our eyes tell us that the loaded hook should be tipping forward and falling, the whole Skyhook system is in balance.

The stability of the Skyhook-belt system is enhanced because the leather belt is somewhat stiff. When a stiff belt is placed in the angled slot of the Skyhook, the belt is unable to hang straight down. The angled slot forces the stiff belt backwards. The long sides of the belt are twisted out and back so that most of the weight of the belt is below and behind the support point (the finger). If a cloth belt is used it is not twisted sufficiently by the angled slot of the Skyhook to move the center of gravity behind the support point. The sides and ends of a cloth belt hang straight down and therefore the weight is below and in front of the support point, and the Skyhook falls.

Assessment

Have students answer the following questions:
1. What is a Skyhook?
2. Why do you think pioneers made toys like this?
3. How does the Skyhook work?

Cross-Curricular Integration

Language arts:
- Have students write stories to answer this question: Why do you think pioneers made toys like this?
- Have students pretend they lived during the pioneer days and sold supplies to the pioneers. One day a friend brings them a new idea for a toy: the Skyhook. They decide to sell the new toy. Have them write a paragraph telling how they would advertise the new toy. Would they rename the Skyhook? How would they describe the Skyhook? Would they decorate the toy? (Remember, they are pioneers and do not have paint.) What would they sell with the Skyhook to make it balance?
- Have students write stories about how they think the Skyhook was invented.
- Read aloud or suggest that students read one or more of the following books:
  ◦ *The Cabin Faced West,* by Jean Fritz (Trumpet Club, ISBN 0440840317)
    Life is very difficult for a girl and her family when they move to the wilderness of western Pennsylvania.
  ◦ *The Courage of Sarah Noble,* by Alice Dalgliesh (Macmillan, ISBN 0-689-71057-7)
    This story is about a pioneer girl who went with her father to build a new home.
  ◦ *If You Sailed on the Mayflower,* by Ann McGovern (Scholastic, ISBN 0590451618)
    This book describes the Mayflower and what it was like being a Pilgrim in early Plymouth.
  ◦ *Sarah Morton's Day,* by Kate Waters (Scholastic, ISBN 0-590-42635-4)
    This book is about a day in the life of a Pilgrim girl.
  ◦ *Toys Today and Yesterday,* by Virginia Arnold (Macmillan, ISBN 0-02-175020-3)
    Discusses toys from now and long ago.
  ◦ *Trouble for Lucy,* by Carla Stevens (Houghton Mifflin, ISBN 0395289718)
    Lucy and her family become pioneers in Oregon.
  ◦ *The White Stallion,* by Elizabeth Shub (Greenwillow, ISBN 0688012108)
    A family makes their way west on a wagon train to the huge new state of Texas.

Life science:
- Discuss different animals' ability to balance themselves (squirrel or cat) or to balance other objects (dolphin or sea lion).

Social studies:
- Use the Skyhook to introduce a unit on American pioneers.

Reference

Hewitt, P.G. *Conceptual Physics;* Addison-Wesley: Menlo Park, CA, 1987; pp 187–201.

Contributors

Mark Beck (activity developer), Indian Meadows Primary School, Ft. Wayne, IN; Teaching Science with TOYS peer mentor.

Lezlie Dearnell, Delshire Elementary School, Cincinnati, OH; Teaching Science with TOYS, 1993–94.

Handout Master

A master for the following handout is provided:

• Balancing Belts—Observation Sheet

Copy as needed for classroom use.

THE SKYHOOK

Balancing Belts—Observation Sheet

1. Can you balance the Skyhook by itself? _____

2. Now try to balance the Skyhook with a belt. Can you balance it? _____

3. Trade your belt with another student who has a different kind of belt. Will the Skyhook balance with this belt? _____

4. What does the Skyhook do when you use a soft cloth belt?

   _____

   _____

   _____

5. What happens when you use a leather belt?

   _____

   _____

   _____

6. If the leather belt has a large, heavy buckle, what do you have to do to balance the hook?

   _____

   _____

   _____

7. What happens if you hang the belt by a string through the slot?

   _____

   _____

   _____

   _____

THE SIX-CENT TOP

*What makes a top keep spinning? Students experiment
with rotational inertia and the variables affecting it.*

A Six-cent Top

GRADE LEVELS

Science activity intended for grades K–6
Cross-curricular integration intended for grades K–3

KEY SCIENCE TOPICS

- mass
- rotational inertia
- variables

STUDENT BACKGROUND

Students should understand the concept of mass. If your students
will be recording data, simple practice in data record-keeping
procedures would be helpful. A prior introduction to the concept
of "variable" would also be helpful, although it can be introduced
here.

KEY PROCESS SKILLS

• communicating	Students record their observations.
• collecting data	Students collect data concerning the factors that affect length of spin.

TIME REQUIRED

Setup	20	minutes
Performance	30–40	minutes
Cleanup	5	minutes

Materials

For "Getting Ready" only
- pencil sharpener
- wire cutters

For the "Procedure"
Part A, per student
- 1 2½-inch (6-centimeter) piece of ¼-inch-diameter dowel rod
- scissors

- 1 3-inch (8-centimeter) square of oaktag or paperboard
These could be cut out of cereal boxes.

- 6 pennies
- 6 small paper clips

Per class
- variety of spinning tops
- hot melt glue gun and glue stick or rubber cement

For older students, per group of 4 students
- a simple stopwatch
- compass
- protractor

For "Variations"

❶ All materials listed for the "Procedure" plus the following:
- 6 washers with center holes large enough to fit over the ¼-inch dowel rod

❷ All materials listed for the "Procedure" except
- substitute 2½-inch squares of oaktag or paperboard

Safety and Disposal

For younger students, gluing with hot glue or rubber cement should be done by an adult. No special disposal procedures are required.

Getting Ready

For the activity, each student will need a 3-inch (8-centimeter) diameter oaktag or paperboard circle. Poke a small starter hole, approximately $^1/_{16}$ inch (2 millimeters) in diameter, in the center of each circle, and make pencil marks every 60 degrees. For younger students, either make these in advance or make a few patterns for the students to use to make their own circles. Cut dowel rods to 2½-inch (6-centimeter) lengths. They cut easily with wire cutters. With a pencil sharpener, sharpen each dowel rod to nearly a point (about 1 millimeter wide) to reduce surface friction. The sharpening must be well centered on the dowel rod. This can be accomplished if you use a sharpener with a variable hole size set for ¼ inch.

Introducing the Activity

Bring out a variety of spinning tops for demonstration and student experimentation. Ask students to describe the various tops according to their size and shape. Find what is common about all of the tops—most likely that they all have a wide middle and narrow to about a 1-millimeter point on the bottom. Ask students how they would design a top of their own.

Part A: Preparing the Tops

1. Give each student a dowel rod. Ask each student to put the point of it down on the floor or table and spin it like a top. "What happens?" *The dowel rod quickly falls over on its side.* Ask, "What does the dowel rod need to work like a spinning top?" *A wider middle or more mass around the middle.* Have students record their observations on the "Spinning Tops" Observation Sheet (provided).

2. If the oaktag or paperboard circles were prepared in "Getting Ready," distribute one to each student. If circle patterns were prepared in "Getting Ready," distribute an oaktag or paperboard square to each student and have the students share the patterns to make their own circles. For older students, give them oaktag or paperboard squares and have them make their own circles using compasses and make pencil marks every 60 degrees using protractors.

3. Have each student poke the point of the dowel rod through the starter hole in the circle without tearing the hole any bigger than necessary. The dowel rod should protrude about ¾ inch. Place a dot or two of glue on each dowel rod to hold it to the circle. (See Figure 1.)

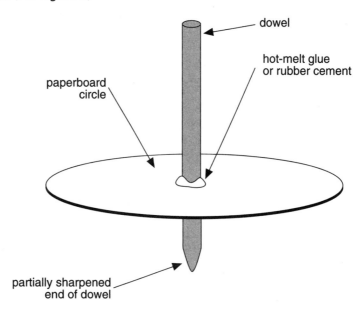

dowel

hot-melt glue or rubber cement

paperboard circle

partially sharpened end of dowel

Figure 1: Construct the Six-Cent Top as shown.

4. Have each student give the dowel rod a spin by twirling it in the fingers like a top. Ask "Is the top any different in behavior now? What made the difference?" Have students practice with the tops a few minutes until they become proficient at getting the top to spin on each attempt. Have them record their observations.

Part B: Experimenting with the Tops

1. (optional) If students are collecting data, each group should choose one group member's top to experiment with. They should now spin the top several times, timing and recording the length of time the top spins before touching its circle on the table.

2. Have students predict what will happen if they clip six paper clips to their tops and record their predictions. Have all students add more mass to their tops by clipping on six paper clips at the pencil marks. Again let students practice until they are proficient. Ask students to make an observation using their sense of touch. "Is it now harder or easier to start the top spinning?" (It should be slightly harder to start to turn the top.) "Does the top have more or less spin time now when you release it?" (It will usually have a longer spin time with this extra mass.)

3. (optional) If students are collecting data, each group should now spin the same top as used in Part B, Step 1, 10 times, timing and recording the length of time the top spins without touching its circle on the table.

4. Have all students remove the six paper clips and hold them in one hand. Have them hold six pennies in the other hand. Ask, "Which feel heavier?" *The pennies feel heavier.*

5. Have students predict what will happen if pennies are glued to the top instead of paper clips and record their predictions. Glue six pennies on each student's circle at the pencil marks. See Figure 1. Let students practice spinning the tops until they are proficient. Repeat the questions asked in Part B, Step 2. (The tops should be much harder to start up but should remain spinning longer.) Have them record their observations.

6. (optional) If students are collecting data, each group should now spin the same top as used in Part B, Steps 1 and 3 10 times, timing and recording the spin time as before.

7. Lead students in a discussion of their observations of the top. Focus the discussion on the variable (mass) and how changes in this variable affected the spinning of the top. Keep a sample of each type of top on hand. Let a student try the different tops if he or she does not remember the feel or performance of the first or second top.

8. Ask students to use their own words to answer the question, "How does mass affect the spinning of a top?"

9. (optional) If students are collecting data, they should now compile data into a usable form by averaging and comparing the average spin times for each variable, then use that data to help answer the question in Part B, Step 8.

Variations

1. Test a different variable (the position of the mass on the top) by using washers with a center hole large enough to fit over the dowel rod. Have students experiment with the mass of six washers stacked in the center and then with the six washers glued near the perimeter.

2. For younger students, make the paper circles 2½ inches in diameter so that the pennies will touch all the way around the circle. (See Figure 2.) This top will be easier for younger students to make.

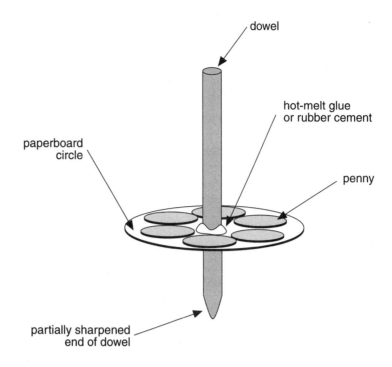

dowel

hot-melt glue
or rubber cement

paperboard
circle

penny

partially sharpened
end of dowel

Figure 2: Construct the Six-Cent Top as shown.

Explanation

 The following explanation is intended for the teacher's information. Modify the explanation for students as required.

Tops are fun to play with because they keep spinning for some time after being set in motion. In this activity, students observed that the duration of the spin is affected by the amount of mass near the edge of the circle. Why? Once in motion, an object (such as a top) that rotates about an axis (in this case, the dowel rod) tends to keep rotating about its axis. The resistance of an object to change its rotational state of motion is called rotational inertia. When the top is spun it tends to keep spinning because of this inertia. When more mass is added near the edge of the circle the rotational inertia is increased. The increased rotational inertia makes it harder to start the top turning, but once it has started turning it will continue to spin longer. Friction eventually causes the top to slow down.

The rotational inertia depends not only on the mass of the spinning top but also on the location of the mass. The farther the mass is located from the axis of rotation the greater the rotational inertia. A top with six washers glued near the perimeter will have more rotational inertia than a top with the same mass but with the washers stacked near the center. Many real machines have the mass located far from the axis of rotation in order to increase the rotational inertia and rotational kinetic energy. For example, at the Ford Museum in Dearborn, Michigan there is a large display of old steam engines. Each one of these engines has a flywheel constructed as a wheel with spokes and most of the mass located near the perimeter.

Cross-Curricular Integration

Language arts:
- Make a class book entitled, "Things that Spin." Have each student contribute an illustrated page to the book that includes an explanation of the illustration.

Math:
- Have the students "purchase" their supplies to make the tops with pretend money.

Reference

Hewitt, P.G. *Conceptual Physics;* Addison-Wesley: Menlo Park, CA, 1987; pp 187–201.

Contributor

Mark Beck (activity developer), Indian Meadows Primary School, Ft. Wayne, IN, Teaching Science with TOYS peer mentor.

Handout Master

A master for the following handout is provided:
- Spinning Tops—Observation Sheet

Copy as needed for classroom use.

Name _____   Date _____

THE SIX-CENT TOP

Spinning Tops—Observation Sheet

1. Describe what happens when you try to spin the stick.

   _____

   _____

   _____

2. What can you do to your stick to make it spin like a top?

   _____

   _____

   _____

3. Describe what happens when you add the cardboard circle to the stick.

   _____

   _____

   _____

4. a. Predict what will happen when you add six paper clips to the cardboard circle.

   _____

   _____

   b. Describe what you observe.

   _____

   _____

5. a. Predict what will happen when you glue six pennies to the cardboard circle.

   _____

   _____

   b. Describe what you observe.

   _____

BOUNCING BALLS

Do all balls bounce alike? Students discover that different balls rebound to different heights.

Balls bouncing to different heights

GRADE LEVELS

Science activity appropriate for grades K–3
Cross-Curricular Integration intended for grades K–3

KEY SCIENCE TOPICS

* energy
* gravity

STUDENT BACKGROUND

Students should have prior experience observing falling objects and understand that gravity makes objects fall. Students should practice measuring heights of ball bounces using a meterstick.

KEY PROCESS SKILLS

• estimating	Students estimate how high a ball has bounced using observations made during several trials.
• making graphs	Students prepare bar graphs to compare elasticity of several types of balls.

TIME REQUIRED

Setup	10	minutes
Performance	30	minutes
Cleanup	5	minutes

Materials

For "Introducing the Activity"
* 6-inch (15-centimeter) rubber band
* padlock

For the "Procedure"

Per group of 3–4 students
- 4 or more of the following types of balls:
  - high-bounce ball (super ball)
  - rubber baseball
  - Ping-Pong™ ball
  - Nerf® ball
  - golf ball
  - tennis ball
  - marble
  - steel ball bearing
- a piece of paper about 1 meter (m) x 1½ m
- masking tape
- meterstick
- different-colored markers
- sticker dots in colors corresponding to marker colors

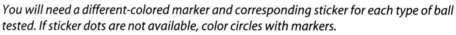

You will need a different-colored marker and corresponding sticker for each type of ball tested. If sticker dots are not available, color circles with markers.

For "Extensions"

❶ All materials listed for the "Procedure" except
Per group
- substitute other types of balls for the ones listed above

❷ All materials listed for the "Procedure" except
Per group
- substitute other types of floor materials for the one used in the "Procedure"

Safety and Disposal

No special safety or disposal procedures are required.

Getting Ready

This activity requires an area for each group near a wall large enough to tape up a piece of paper about 1 m x 1½ m.

1. Attach a large piece of paper to the wall for each group. The bottom edge of the paper should touch the floor.

2. Using a meterstick, measure 1 m above the floor and draw a line across the paper at this height.

3. Place one of each color sticker dot on the line (See Figure 1) and write the name of the ball to be dropped under each sticker.

4. Use color sticker dots to mark the balls. Be sure that the colors used correspond to those on the charts.

Introducing the Activity

Demonstrate the concept of elastic potential energy (stored energy) using a rubber band and a padlock by doing the following: Lock the padlock through the rubber band. While holding the rubber band in one hand, twist the lock. This will wind up the rubber band. It is not necessary to wind the rubber band tight; a few turns are sufficient depending on the type of rubber band. Let go of the lock. The rapid untwisting of the lock and rubber band demonstrates the release of stored energy. The rubber band unwinds itself using the energy you have put into the system.

Procedure

1. Divide the students into groups of three or four and assign jobs.

   Each group will need a Recorder, a Dropper, and one or two Watchers to perform the following jobs:
   - *Recorder—to record the group's predicted and actual results;*
   - *Dropper—to drop the balls; and*
   - *Watcher—to watch the balls bounce and show the Recorder where to mark the bounce.*

2. Give each group a set of balls color-coded with sticker dots. Instruct students not to bounce, throw, or roll any ball until they are told to do so.

3. Have each group predict how high each ball will bounce when dropped from 1 m. Have each Recorder mark the group's predictions on the chart with a pencil. (See Figure 1.)

4. For each ball, have each Dropper hold the ball in the same position as the sticker dot on the chart, with the bottom of the ball parallel to the 1-m line. Have the Droppers drop the balls from this height five to 10 times.

5. Before the ball is dropped, have the Watchers in each group sit on the floor below the ball. They will record how high the ball bounces. They do this first by watching the ball bounce several times to get an idea of how high it will go. After the last bounce, they should position their fingers on the paper at the height where the ball bounces. Once all members of the group agree that this is the correct height for the bounce of the ball, have the Recorder mark this spot on the paper with a colored marker corresponding to the sticker-dot color on the ball.

6. Have the groups repeat Steps 4 and 5 for each type of ball used.

7. Have each group use a meterstick as a straightedge to make a bar graph from the colored marks. The height of the bar corresponds to how high the ball bounced. (See Figure 1.)

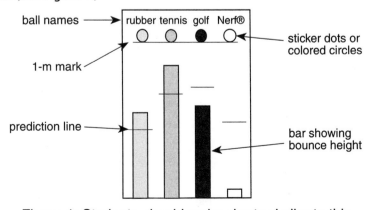

Figure 1: Students should make charts similar to this.

Extensions

1. Use other types of balls.

2. Use only one type of ball, but change the floor material; for example, use a foam pad, a rug, concrete, wood, or steel. Refer to the Teaching Science with TOYS activity "Bounceability."

Class Discussion

Discuss the results (predictions versus actual distances) by asking these questions:

1. "What made the balls fall down?"

2. "Which ball did you predict would bounce the highest? …the lowest?"

3. "Were your predictions correct?"

4. "List balls from highest bouncer to lowest bouncer."

5. Pass around the super ball and Nerf® ball to the students, so they can feel the texture of each. Ask, "How does the super ball feel? How does the Nerf® ball feel?"

6. Ask, "Can anyone tell me why the super ball bounced higher than the Nerf® ball?" After hearing students' ideas, present a simple explanation: Use a stretched rubber band or squeeze a rubber ball to show the elastic properties of the material. Let the rubber band or ball return to its original form. Explain that elasticity gives the super ball its bounce. See the "Explanation" for more information.

Explanation

The following explanation is intended for the teacher's information. Modify the explanation for students as required.

Objects can store energy in many ways. A stretched or twisted rubber band stores elastic potential energy. The stored chemical energy in foods and fuels is a form of potential energy. In this activity, your students explore a form of stored energy called gravitational potential energy.

A ball held at some distance above the ground possesses gravitational potential energy. The force needed to lift the object against gravity is the weight of the object, and work is done lifting it through a distance above the ground. The product of the object's weight and the distance lifted is known as the gravitational potential energy.

Thus, just before a ball is dropped it possesses gravitational potential energy. As it falls, it gains kinetic energy (energy of motion). When the ball collides with the floor, some of this kinetic energy is stored as elastic potential energy in the ball and the floor. The particles in the ball and the floor squeeze together like tiny springs. Because the material in the super ball springs back to its original shape after being deformed, more of the elastic potential energy is returned to the ball, causing it to rebound higher. The collision is said to be elastic. On the other hand, the type of material in a Nerf® ball or clay ball causes it to absorb this potential energy and to return to its original shape slowly or not at all. Much of the energy is not returned to the motion of the ball, resulting in a low bounce. The collision is said to be inelastic.

The "hard" materials like super balls, golf balls, and steel balls are elastic materials—materials that spring back to their original shape after being deformed. Even though these balls feel like they cannot be squeezed, they do compress when they hit the floor. The "soft" materials like a Nerf® ball and a ball of clay are inelastic materials—materials that return to their original shape slowly or not at all after being deformed.

Cross-Curricular Integration

Language arts:
- Read aloud or suggest that students read one or more of the following books:
  - *The Ball Bounced,* by Nancy Tafuri (Greenwillow, ISBN 0-688-07871-0)
    A bouncing ball causes much excitement around the house.
  - *Gravity Is a Mystery,* by Franklyn Branley (Trophy, ISBN 0-06-445057-0)
    Explains in simple text and illustrations what is known about the force of gravity.
  - *Here Comes Alex Pumpernickle,* by Fernando Krahn (Little, Brown, ISBN 0316503118)
    This is a wordless book of a boy's series of misadventures. One episode depicts his ball bouncing into a pot of something sticky.
  - *The Mystery of the Magic Green Ball,* by Steven Kellogg (Dial, ISBN 0803762151)
    A boy named Timothy searches for his ball.
  - *Play Ball,* by Margaret Hillert (Follett, ISBN 0-695-40879-8)
    In this easy reader book, two boys try to find the right ball and equipment to play together.
  - *Stop that Ball,* by Mike McClintock (Random House, ISBN 039490013)
    A little boy hits his tether ball so hard that the string breaks and away goes the ball, leading him to adventure.
  - *Why Doesn't the Earth Fall Up? And Other Not Such Dumb Questions about Motion,* by Vicki Cobb (Lodestar, ISBN 0-525-67253-2)
    Answers nine questions about motion, explaining Newton's Laws of Motion, gravity, centrifugal force, and other principles of movement.

Math:
- Students graph their results.

References

Hewitt, P. *Conceptual Physics;* Little, Brown: Boston, 1985; pp 80–86, 171–172.

Liem, T. "How High Will the Ball Bounce?" *Invitations to Inquiry—Supplement to 1st and 2nd Editions;* Ginn: Lexington, MA; p 137.

Contributors

Sally Drabenstott, Sacred Heart, Fairfield, OH; Teaching Science with TOYS peer writer.

Shirley Haeuptle, Beverly Gardens School, Dayton, OH; Teaching Science with TOYS, 1992–93.

Karen Mitchell, Sherman Elementary School, Toledo, OH; Teaching Science with TOYS, 1992–93.

SNAP, CRACKLE, POP

Rice Krispies® jump in the air as students investigate static electricity.

Static electricity in action

GRADE LEVELS

Science activity appropriate for grades K–1
Cross-Curricular Integration intended for grades K–1

KEY SCIENCE TOPICS

- attractive and repulsive forces
- static electricity

KEY PROCESS SKILLS

• observing	Students observe the effects of static electricity.
• investigating	Students investigate attraction between a charged balloon and rice cereal.

TIME REQUIRED

Setup	15	minutes
Performance	15	minutes
Cleanup	5	minutes

Materials

For the "Procedure"
Per student or learning center
- paper plate
- balloon
- small paper cup of oven-toasted rice cereal, such as Rice Krispies®
- (optional) 1 piece of wool cloth
- (optional) small pieces of paper

Per class
- several hand mirrors

For "Variations and Extensions"
❶ Per class
- access to carpet and metal doorknob

❷ Per class
- 30-centimeter (cm) piece of string
- 2 balloons

❸ Per group
- tissue paper
- clear plastic (deli) container with lid

Safety and Disposal

For health purposes, be sure that two students do not rub their hair on the same balloon. No special disposal procedures are required.

Getting Ready

Inflate balloons.

Introducing the Activity

Begin by asking your students about electricity. They will probably tell you what electricity does: lights light bulbs, makes toasters work, etc. Talk with them about a different kind of electricity: static electricity. This is the kind of electricity they may have experienced during the winter when they got shocked taking off a sweater or walking across a carpet and touching a door knob or another person.

Procedure

 This activity must be performed when humidity is low.

1. Give each student an inflated balloon, a paper plate, a cup of rice cereal, and if desired, a piece of wool cloth.

2. Have the students rub the balloons in their hair or with the piece of wool cloth. Explain that they are charging their balloons with electricity.

3. Tell them to listen carefully. Ask, "What sounds do you hear coming from your hair?" *Popping and crackling sounds.*

 At this point the students may ask if the snap, crackle, pop sounds they hear when they pour milk on rice cereal are caused by static electricity. The answer is no. The air rushes out of the holes in the cereal when milk is poured, causing the crackling sounds. The crackling they heard in their hair is "mini-thunder," which is similar to the physics of the thunder they hear in a storm, but on a much smaller scale.

4. Have students repeat Step 2 to recharge their balloons. Then have them hold the balloons a few centimeters above their hair. Hold up a mirror so they can see the results. Ask, "What happened?" *Our hair stands up.* Tell them that static electricity caused their hair to stand up.

 If the charged balloon will not attract their hair, have them bring the charged balloon near some small pieces of paper. Hair needs to be clean and dry. Naturally oily hair, hair spray, shampoo, or conditioner might prevent a person's hair from becoming charged by absorbing the charge (the electrons) as it is produced during the friction process.

5. Have students pour their rice cereal on their paper plates. Again have them recharge their balloons, then hold their balloons a few centimeters above the plate. Ask, "What happened?" *Some rice cereal pieces jump up and fall down and some stick to the balloon.* Ask, "What caused this?" (Students should be able to answer "static electricity.")

6. Ask, "How many of you have ever sorted socks that have just come from the dryer? Did the socks ever stick together?" Explain that static electricity causes socks to stick together after being in the dryer. Use other common examples such as taking off a sweater.

Extensions

1. In the winter, students can shuffle across the carpet and touch a metal doorknob. They may feel a shock or may see a spark of electricity.

2. Attach a 30-cm piece of string to two balloons. Charge each balloon on the same person's hair. While holding the middle of the string, bring the balloons near each other. They should repel. Have the students try this experiment with both balloons uncharged and with one charged and one uncharged.

3. Have students cut paper tissue into small seasonal shapes (winter, snowflakes; fall, leaves; spring, flowers). Have them place the shapes in a clear plastic (deli) container with a lid. Tell students that they can make the objects dance without touching the container. Have the students hold balloons above the lids and observe. Then have the students rub the balloons on their heads, hold the balloons over the lids, and watch the shapes dance. Challenge the students to explain what happened by drawing a picture.

4. Have students do the "Balloon Pickup" Take-Home Activity (handout provided) outside of school with an adult partner.

Explanation

The following explanation is intended for the teacher's information. Modify the explanation for students as required.

While a detailed explanation of static electricity is beyond the comprehension of kindergartners and first-graders, they can learn some of the effects of static electricity by doing this activity. Students experience the phenomenon that static electricity can make some objects attract each other and others repel each other. The students should not be expected to understand why this attraction and repulsion happens. For an explanation of static electricity, refer to the Teaching Science with TOYS activity "Magic Balloon."

Cross-Curricular Integration

Language arts:
- Have the students write about their discoveries in their science journals.

References

Barhydt, F. *Science Discovery Activities Kit;* The Center for Applied Research in Education: West Nyack, NY, 1982.

Hewitt, P. *Conceptual Physics;* Little, Brown: Boston, 1985; pp 322–335.

Paulu, N. *Helping Your Child Learn Science;* U.S. Department of Education: Washington, D.C., June 1991.

Sarquis, M.; Kibbey, B.; Smyth, E. *Science Activities for Elementary Classrooms;* Flinn Scientific: Batavia, IL, 1989.

Tolman, M.; Morton, J. *Physical Science Activities for Grades 2–8;* West Nyack, NY, Parker: 1986.

Contributors

Sally Drabenstott, Sacred Heart, Fairfield, OH; Teaching Science with TOYS peer writer.

Judy Piehowicz, Fairfield Elementary School, Pickerington, OH; Teaching Science with TOYS, 1993–94.

Jill Stelzer, Murlin Heights Elementary School, Dayton, OH; Teaching Science with TOYS, 1993–94.

Handout Master

A master for the following handout is provided:

- Balloon Pickup—Take-Home Activity

Copy as needed for classroom use.

SNAP, CRACKLE, POP

Balloon Pickup—Take-Home Activity

Date _____

Dear Adult Partner:

We have been experimenting with static electricity and balloons. We talked about how things attract because positives and negatives stick together. As we learned with magnets, the north pole and the south pole pull together, and static electricity is similar. We explored this principle in class by using dry substances and balloons. The students rubbed balloons on their hair and then put the balloons close to the substances. They watched the pieces jump up and down near the balloons.

I would like you to try this activity at home. You may use any items you choose; some suggestions are flour, sugar, Cheerios®, parsley flakes, cracker crumbs, or potato chip crumbs. Please write the names of what you tried on the take-home record sheet. Circle what happened when the balloon came near the items after you charged it by rubbing it on your hair. I would appreciate any comments you or your child have about this activity. Please return this record sheet by _____. Thank you and have fun experimenting.

Sincerely,

_____

SNAP, CRACKLE, POP

Balloon Pickup—Take-Home Activity, page 2

Test several common items in your home to see if there is any static electricity between these items and a balloon. Circle the appropriate picture to describe what happened. Please write comments on the back of this sheet.

Attracted   **No Attraction**

1.

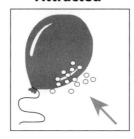 

2.

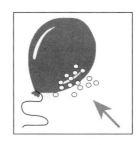 

3.

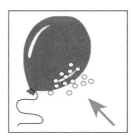 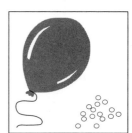

4.

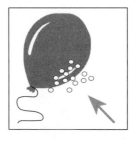 

5.

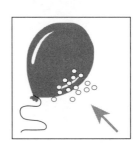 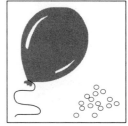

Teaching Physics with TOYS

MAGIC BALLOON

Look out, Houdini! In this activity, student "magicians" use Magic Balloons to make objects jump.

A "Magic Balloon" attracting puffed rice

GRADE LEVELS

Science activity appropriate for grades K–3
Cross-Curricular Integration intended for grades K–3

KEY SCIENCE TOPICS

- attractive and repulsive forces
- positive and negative charges
- static electricity

KEY PROCESS SKILLS

• collecting data	Students collect data concerning the nature of positive and negative charges.
• investigating	Students use a charged balloon to investigate static electricity.

TIME REQUIRED

Setup	20	minutes
Performance	20	minutes
Cleanup	10	minutes

Materials

For the "Procedure"
Per pair of students
- 2 plastic straws
- 2 balloons
- 2 single-ply sheets of facial tissue
- small paper cup
- 3 ounces puffed rice
- small paper plate
- (optional) tape or string
- (optional) wool cloth

Per class
- salt and pepper shaker

For the "Variations and Extensions"

❶ All materials listed for the "Procedure" plus the following:
Per class
- Ping-Pong™ ball

❷ All materials listed for the "Procedure" plus the following:
Per class
- roll of ¾-inch Scotch® brand Magic™ Tape

Safety and Disposal

For health purposes, be sure that two students do not rub their hair on the same balloon. Caution students not to eat the pepper or get it in their eyes. No special disposal procedures are required.

Getting Ready

Inflate the balloons and attach one balloon to each straw with tape or string, or just tie the end of the balloon around the straw to make a Magic Balloon.

Introducing the Activity

Ask, "Have you ever seen a balloon stick to someone's head? How do you make it stick?" *Rub it on someone's hair.* Demonstrate by rubbing a balloon on your hair. This will put a negative charge on the balloon. Explain that when some objects rub together, negative charges move from one to the other. Bring the charged balloon near your hair; your hair and the balloon will attract. Explain that when the negative charges left your hair, your hair became positively charged. Since negative and positive charges attract, the balloon and your hair stuck together. Now, charge two balloons by rubbing on your hair. Ask, "Are these balloons negatively or positively charged?" *They are negatively charged.* "What will happen when they come near each other?" *They will push each other away.* Demonstrate for the students.

The nature of a material determines what charge it will acquire when rubbed against another material. For example, when you rub a balloon in your hair, the balloon becomes negatively charged because rubber and plastic materials have a stronger tendency to become negatively charged than hair does. A glass rod rubbed with silk or a piece of acetate rubbed with cotton will become positively charged because silk and cotton have a stronger tendency to become negatively charged than glass or acetate.

Procedure

1. Have one student in each pair tear tissue into little pieces and put them on the plate while the other gently rubs the balloon on his or her hair.

   A piece of wool cloth may be used as an alternative to hair. Be sure that each student in each pair has a balloon and that only one student rubs his or her hair on any one balloon.

2. Explain to the students that they have now charged their balloons—some negative charges are on the surface of the balloon.

3. Have students in each pair take turns holding the charged balloon over the plate. Ask, "What happened?" *The tissue stuck to the balloon.* Ask, "If the balloon has negative charges on its surface, then what charges must be on the surface of the tissue?" *Positive.* Ask, "Where did the negative charges come from?" *The hair or wool.* Remind students that objects with positive charges and objects with negative charges come together. Explain that they have seen static electricity in action.

4. Have the students dispose of the tissue pieces. Have one student in each pair pour the puffed rice on the plate. Have the other student rub the balloon on his or her hair and wave the charged "Magic Balloon" over the puffed rice. Ask, "What happened?" *The charged balloon picks up the pieces of rice. The negative charges on the balloon attract the positive charges in the rice.*

5. Have the students put the puffed rice back into the cups, and then sprinkle equal amounts of salt and pepper on the plate.

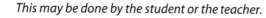
This may be done by the student or the teacher.

6. Challenge the students to figure out how to separate the salt and pepper by using static electricity. Have one student in each pair rub the balloon on his or her hair or a piece of wool and hold the Magic Balloon high over the mixture. Ask, "What happened?" The salt should stay on the plate while the balloon picks up the pepper.

7. Ask, "What other examples of static electricity have you experienced?" Possible answers include static cling on clothes coming out of the dryer, hair standing up after being combed, being shocked after taking off a sweater, or walking across a carpeted room and being shocked when reaching for a metal doorknob or another person.

Extensions

1. Second- and third-grade teachers can extend the experiment by charging the balloon and placing it near a Ping-Pong™ ball. The Ping-Pong™ ball will roll toward the balloon.

2. Have the students use two pieces of Scotch® brand Magic™ Tape to create two oppositely charged objects. Tear off two 15-centimeter pieces of tape. Fold tabs on both ends of each piece of tape. Stick one piece of tape to a clean area on the desk. Write the letter "B" (for bottom piece) on the tab of this piece of tape. Stick the other piece of tape directly on top of the piece of tape marked "B." Write the letter "T" (for top piece) on the tab of this piece of tape. Holding only the bottom tab, pull the tapes off the desk. The two pieces of tape are now stuck together, sticky side to non-sticky side. Holding the tabs "B" and "T," pull the two pieces of tape apart rapidly. One piece of tape will be attracted to the balloon. The other one will be repelled.

Explanation

 The following explanation is intended for the teacher's information. Modify the explanation for students as required.

All matter is made up of small particles. These particles are made up of parts that have electric charges—protons that have a positive charge and electrons that have a negative charge. The structure of these particles sometimes allows some of the electrons to be removed. For example, when rubbing a balloon against your hair, the friction causes some of the electrons from your hair to be transferred to the balloon. After this exchange, the balloon becomes negatively charged with an excess of electrons, and the hair is positively charged with a shortage of electrons.

Objects with the same net charge repel each other; for example, two negatively charged objects repel, as do two positively charged objects. After rubbing your hair with a balloon, this repulsion causes the positively-charged strands of hair to repel each other and therefore stand on end. Objects with opposite charges attract each other. If the students have studied magnets, explain that electric charges behave the same as magnetic poles. (Like poles repel and opposite poles attract.)

Objects can also attract each other when one is negatively charged and the other is not charged. This happens because negatively charged objects (such as a charged balloon) can repel the negative charge on another object, causing these negative charges to move to another area on the object. When the negative charges move away, a positive region is created which is then attracted to the original negatively charged object. This is what happens to the tissue paper, rice puffs, and pepper. The pepper is lifted by the balloon because the electrical attraction force is greater than the weight of the pepper particles. The salt crystals are larger and heavier, so the electrical force may or may not be large enough to lift them.

Cross-Curricular Integration

Language arts:
- Have students write a "how-to" paragraph about one of the static electricity activities completed in class. Have them list the materials and write sentences listing the steps in order to complete the activity. Assess the paragraphs by having students try to complete the activity based on the how-to paragraphs. Have students make a class book of the paragraphs.
- Have students write letters, stories, poems, and paragraphs about static electricity.
- Have students write a concrete poem about static electricity using a balloon as the concrete object with words describing static electricity written around the outline of the balloon. (See Figure 1.)

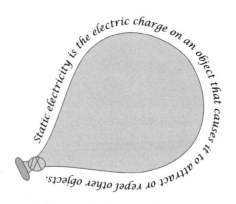

Static electricity is the electric charge on an object that causes it to attract or repel other objects.

Figure 1: Have students write concrete poems.

Home, safety, and career:
• Have students go to a store with an adult partner and list static-guard products sold in the store. Discuss when these products would be used.

References

Hewitt, P. *Conceptual Physics;* Little, Brown: Boston, 1985; pp 322–335.

Paulu, N. *Helping Your Child Learn Science;* U.S. Department of Education: Washington, D.C., 1991.

Sarquis, A.; Kibbey, B.; Smyth, E. *Science Activities for Elementary Classrooms;* Flinn Scientific: Batavia, IL, 1989.

Tolman, M.; Morton, J. *Physical Science Activities for Grades 2–8;* Parker: West Nyack, NY, 1986.

Contributors

Lezlie Dearnell, Delshire Elementary School, Cincinnati, OH; Teaching Science with TOYS, 1993–94.

Sally Drabenstott, Sacred Heart, Fairfield, OH; Teaching Science with TOYS peer writer.

MAGNET CARS

How do magnets change the way toy cars act? What if we had magnets on real cars?

A Lego® car with magnet attached

GRADE LEVELS

Science activity appropriate for grades K–3
Cross-Curricular Integration intended for grades K–3

KEY SCIENCE TOPICS

- attractive and repulsive forces
- magnets
- north and south poles

STUDENT BACKGROUND

Young students will probably understand the terms "push away" and "pull towards." This activity will supply the experience necessary to apply these terms to magnets.

KEY PROCESS SKILLS

- observing — Students observe the behavior of magnet cars compared to cars that have no magnets attached.

- defining operationally — Students define magnetism in terms of "push" or "pull" of one object on another.

TIME REQUIRED

Setup	10	minutes
Performance	30	minutes
Cleanup	10	minutes

Materials

For the "Procedure"
Per student or group of 3 students:
- Lego® construction block sets (enough to build 3 cars) or at least 3 Lego® single car kits

 Hot Wheels® or Matchbox® cars can be used in place of Lego® cars.

- 2 rubber bands, small pieces of Velcro™, or adhesive tape
- 2 disk magnets (recommended) or bar magnets (alternative)

Safety and Disposal

No special safety or disposal procedures are required.

Procedure

1. Place canisters of Lego® construction sets or single car sets on the floor. Have each student make three toy cars. Give the students time to build and play with the cars.

   ➤ *If enough parts are not available, divide students into groups of three and have each student make one car.*

2. After the free play time, use rubber bands to attach either a bar magnet to the tops or a disk magnet to the fronts, backs, or tops of two of each student's (or group's) cars. (See Figure 1.) Do not attach a magnet to the third car. Be sure some cars have magnetic north facing forward and some have magnetic south facing forward.

   ➤ *Test that some cars have north facing forward and some have south facing forward by putting cars together and making sure some attract each other and some repel each other.*

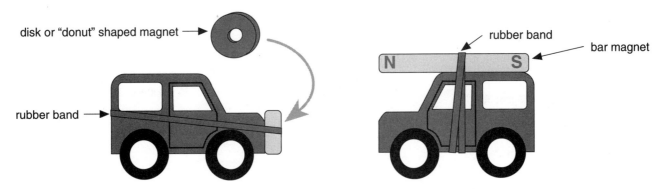

disk or "donut" shaped magnet →

rubber band →

rubber band

bar magnet

N   S

Figure 1: Attach the magnets to the toy cars.

3. Tell the students that two of their cars now have magnets on them and allow them to play with all the cars again.

4. After the students have played with the cars, ask the following questions:

   a. "Do the magnet cars do things your plain cars did not do? How do magnets change the way the cars act?" *Some cars stick together and some push each other away.*

   b. "What happens if a magnet car and a plain car get close together?" *Nothing.*

5. Summarize the students' observations.

   • Cars without magnets will always travel in a straight line when pushed and will not be pulled aside when near a car with a magnet.

   • Two cars with magnets will either roll toward each other and stick together (north attracts south) or push away from each other (north repels north and south repels south).

6. Have second- and third-graders learn the terms "repel" and "attract."

Variation

Because kindergarten students have short attention spans, kindergarten teachers may wish to present this activity in two sessions. In the first session the students could make their cars. In the second session they could work with the magnet cars and the plain car.

Explanation

 The following explanation is intended for the teacher's information. Modify the explanation for students as required.

For young students an explanation of magnetism using the terms "pull towards" and "attract" when opposite poles interact and "push away" and "repel" when like poles interact is sufficient.

Magnetism arises from the motion of negative electric charges (electrons). Electrons produce magnetic fields through two types of motion: first, the electrons moving about the nucleus of the atom; and second, the electrons spinning like tops as they move. Because all atoms contain electrons, their motions cause every atom to act like a tiny magnet.

Materials are said to be magnetic when the magnetic fields of the atoms all line up in the same direction. In a substance like iron, nickel, or cobalt, the atoms line up in groups creating a magnetic domain. Each domain is made up of millions of aligned atoms. If many domains within a substance are lined up, the magnet is strong. If the domains are not lined up, the magnet is weak.

Like electric charges, magnets exhibit both attractive and repulsive forces. Simple bar magnets are labeled with an "N" for north pole and "S" for south pole to designate the opposite magnetic properties. The terms "north pole" and "south pole" come from the observation that a sample of magnetic material suspended from a string will align itself with the Earth's magnetic field. One end points toward the Earth's geographic north pole, and the other end points toward the Earth's geographic south pole. The nature of magnetic poles is such that like magnetic poles "push away" from each other and opposite magnetic poles "pull toward" each other.

Cross-Curricular Integration

Language arts:
- Have the students write a story about where their cars are going.

Social studies:
- Review the four compass directions. Talk about two magnet cars repelling each other going in opposite directions: Ask, "If Car A is going east, in what direction will Car B go? and vice versa?" Lay the cars down on a map of the U.S. Put the cars on an east-west line in the middle of the country. Ask, "If these cars are repelling each other and one is heading toward the Atlantic Ocean, toward what ocean is the other one heading?" Or put the cars on a north-south line in Colorado and ask, "If one car is heading toward Canada, toward what country is the other car headed?"

References

Althouse, R.; Main, C. *Science Experiences for Young Children, Magnets;* Teacher College: New York, 1975; pp 9–11.

Hewitt, P. *Conceptual Physics,* 5th ed.; Little, Brown: Boston, 1985; pp 125–143.

Contributor

Sally Drabenstott (activity developer), Sacred Heart, Fairfield, OH; Teaching Science with TOYS peer writer.

SCHOOL BOX GUITAR

Let's rock-n-roll! Students use school supplies to make music and explore sound.

A School Box Guitar

GRADE LEVELS

Science activity appropriate for grades K–3
Cross-Curricular Integration intended for grades K–3

KEY SCIENCE TOPICS

- longitudinal wave
- sound
- vibration

KEY PROCESS SKILL

- controlling variables   Students control the pitch of sound by varying tension or type of band.

TIME REQUIRED

Setup	10	minutes
Performance	20	minutes
Cleanup	5	minutes

Materials

For the "Procedure"
Part A, per student
- 1 12-inch or 30-centimeter ruler

Part B, per student
- 1 cardboard school box or other similar box
- 4–5 rubber bands with different lengths and widths
- 1 12-inch or 30-centimeter ruler

Part C, per class
- Slinky®

For "Variations and Extensions"
❶ Per class
- various musical instruments

❷ Per class
- unwound clock or unwound music box

Safety and Disposal

No special safety or disposal procedures are required.

Getting Ready

Make a sample School Box Guitar by stretching four or five different-sized rubber bands lengthwise over the opened box.

Procedure

Part A: Sounds Produced with Different Lengths of Ruler

1. Place a 12-inch ruler flat on a table. Let 4 inches of the ruler extend over the edge of the table.

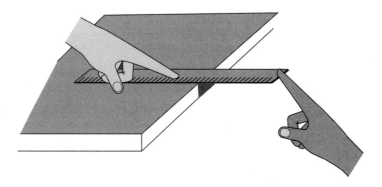

Figure 1: Pluck the free end of the ruler.

2. Hold the ruler securely to the table with one hand and pluck the end of the ruler that sticks out over the end of the table with the other hand.

3. Have students duplicate Steps 1 and 2 with their own rulers.

4. Ask students, "What did you see? What did you hear?" *The ruler vibrates up and down. A sound is heard.*

5. Have the students experiment with different lengths of ruler extending over the end of the table. Point out that they are making different sounds.

Part B: Sounds Produced with Different Rubber Bands

1. Bring out the sample School Box Guitar. Pluck the rubber bands one at a time to make different sounds.

2. Have students make their own School Box Guitars and pluck the rubber bands.

3. Ask students, "Why are the sounds of the various bands different?" *The rubber bands have different masses and are under different amounts of tension. These differences contribute to different frequencies of vibration and therefore different sounds.*

4. Explain to students that the energy from a vibrating object transfers to gas particles, which pass their energy to adjacent particles, producing a longitudinal wave.

Part C: Making Longitudinal Waves with a Slinky®

1. Demonstrate a longitudinal wave as follows: While holding one end of a Slinky® fixed on the floor or a flat table, move the other end back and forth. (See Figure 2.)

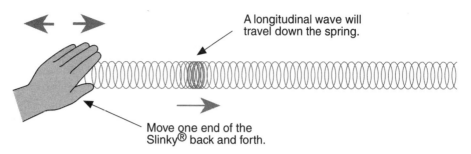

A longitudinal wave will travel down the spring.

Move one end of the Slinky® back and forth.

Figure 2: Use a Slinky® to show the propagation of longitudinal waves.

Extensions

1. Have students apply the concepts learned with the School Box Guitar to other string instruments or even to percussion instruments.

2. Pass around an unwound clock or unwound music box. Ask students, "Is this object producing any vibrations?" Wind up the clock or music box and pass it around again. Students can compare what they hear and feel from the vibrations produced.

3. Discuss how the ear receives sound. A diagram or model of the ear is necessary with students of this age. The eardrum can be compared with an instrument—the drum. Instead of using the more difficult terms, "anvil" (incus), "hammer" (malleus), and "stirrup" (stapes) can be used to name the bones in the ear.

Explanation

The following explanation is intended for the teacher's information. Modify the explanation for students as required.

In this activity, sound is produced when the energy from the vibrating ruler/rubber band gets transferred to gas particles in the air that are moving freely and happen to collide with the ruler or rubber band. These gas particles pass their energy to adjacent particles, producing a longitudinal wave. A longitudinal wave is a wave in which the particles of the material vibrate in the same direction that the wave disturbance travels.

Different sounds are produced by different rates of vibration. The rate of vibration is known as the frequency. Higher-pitched sounds are produced by faster rates of vibration, in other words, by higher frequencies. Lower-pitched sounds are produced by slower rates of vibration, in other words, lower frequencies. The school box acts as a sounding board to amplify the vibrations produced by the vibrating rubber bands. The rubber bands have different masses and are under different amounts of tension. These differences contribute to different frequencies of vibration and therefore different sounds.

Cross-Curricular Integration

Art:
- Have students design and make their own instruments.

Language arts:
- Read aloud or suggest that students read the following book:
  - *Whistle for Willie*, by Ezra Jack Keats (Bodley Head, ISBN 0370007603)
    A story about how to pucker your lips to produce vibration and sound.

Life science:
- Have students study how the ear works and what can cause hearing loss.

Music:
- Have students study different types of instruments and discuss how they produce vibrations.
- Have students investigate the musical instruments created by different cultures in different historical periods.

References

Harlan, J. *Science Experiments for the Early Childhood Years,* 4th ed.; Merrill: Columbus, OH, 1988; pp 227–228.

Hewitt, P. *Conceptual Physics,* 5th ed.; Little, Brown: Boston, 1985; pp 280–305.

Contributor

Sally Drabenstott (activity developer), Sacred Heart, Fairfield, OH; Teaching Science with TOYS peer writer.

Activities for Grades 4–6

FORCES AND MOTION

Students discover some of the basic principles governing motion.

Magnetic race car with magnetic wand

KEY SCIENCE TOPICS

- forces
- Newton's laws of motion
- friction
- gravity
- magnetism

STUDENT BACKGROUND

This lesson will probably be used as an introduction, so no previous knowledge of forces is necessary. However, it may alternatively be used as a review lesson for a study of forces and motion.

KEY PROCESS SKILLS

• observing	Students observe the action of a series of forces (gravity, magnetism, buoyancy, etc.) on objects.
• hypothesizing	Students hypothesize how changes in force will affect motion.

TIME REQUIRED

Setup		negligible
Performance	40–60	minutes
Cleanup		negligible

Materials

Per class:
- plastic car
- small magnet
- paper clips
- masking tape
- water rocket or balloon

- magnetic race car

The race car contains a concealed magnet; thus, it can be pushed or pulled by the accompanying magnetic wand.

- wooden block or any object that will float
- container of water large enough to submerge wooden block
- 2 similar-sized balls with different masses (for example, a basketball and a cheap, inflatable ball or a whiffle ball and a softball)
- a large book or a meterstick
- (optional) bowling ball
- a rough surface such as a strip of carpet or sand on the floor or several strips of sandpaper
- a smooth surface such as posterboard or a sheet of scrap linoleum

Safety and Disposal

The water rocket should be set off in a safe place and manner. Outdoors is best, but a gym with a high ceiling will do. If used in a classroom, the rocket should be used with very little water. No special disposal procedures are required.

Getting Ready

Practice setting off the water rocket. Make a starting line on the floor to use in measuring the distance the car traveled.

Procedure

You and the materials should be located where everyone can see the floor. This might mean being in the center of the room, on a raised stage, or in a small group. Weather permitting, doing the lesson outside might also work well. Have students record their responses to discussion questions on the "Discussion Record" (provided).

1. Push the car across the floor. Ask the students, "What made the car move?" *Your push made the car move.* After you and the students establish that, ask them how else you might make the car move. Possible responses are: *Blow on it, set it on a hill or ramp and let it go, run another car into it.* Ideally you would stop here and try out any suggested methods.

2. Ask students what all these ways of making the car move have in common. *In each case, a force is exerted on the car.* (Introduce the term force here if your students don't know it.) Tell the students that nothing ever starts moving by itself. Whenever something starts moving, something has made it move. The thing that made it move is a force. A force is a push or a pull. You may wish to write on a poster or the board, "When something moves, a force made it move."

3. In each of the following cases, demonstrate and ask students if they can name the force (push or pull):

   a. Drop a ball. *Gravity.*

   b. Use a magnet to pick up paper clips. *Magnetic force.* Ask if there is any way you could use magnetic force to make a toy car move. You cannot do it directly, but you can tape a magnet to the bumper and then use a magnetic wand or other magnet to push or pull. Demonstrate the magnetic race car that already has a magnet inside the car.

c. Set off the water rocket. (You might call this *reaction force*. The compressed air in the rocket pushes the water down. The water then exerts an upward reaction on the compressed air and the rocket. This is not a complete explanation, but it is sufficient for now.) A water rocket set off outdoors makes an exciting demonstration. However, if a water rocket or a suitable launch area is not available, a balloon blown up and released can illustrate the same principle.

d. Place under water a wooden block or any object that will float and release it. *Buoyant force.* If you release a piece of wood underwater, it will rise. The pressure in a liquid increases with depth; therefore, the water pushes up on the bottom of the block with a larger force than it pushes down on the top of the block. For an object such as the wood block, which floats, this net upward force is greater than the object's weight, so the object rises. Objects sink if their weight is larger than the net upward force of the water.

4. Push the plastic car again, or better yet, have a student do this. Ask what determines how far the car travels. List responses on the board, adding, when they're finished, any of the following that they've not thought of:

   a. how hard you push it;

   b. how long you push it;

   c. how smooth the surface is; and

   d. how heavy the car is (its mass).

5. Discuss and demonstrate each of the factors listed in Step 4.

   a. Push the car twice, making a clear difference in how hard you push. You and the students should observe that the bigger the force, the farther the car travels.

   b. Push the car twice, keeping your hand on the car longer the second time. Try to use equal forces. You should observe that the longer the force acts, the farther it travels. Point out that this is one of the reasons for follow-through in sports such as tennis or kickball.

   c. Use two surfaces—a rougher one and a smoother one. Run the car on each of the two surfaces, pushing it as similarly as possible. You should observe that the car will go further on the smoother surface. The smoother the surface, the smaller the friction force between the surface and the object. A possible extension is to discuss what would happen on a surface with absolutely no friction whatsoever (if this were possible) and then make the connection to the way things move in space.

   d. Use two balls of about the same diameter having much different masses, such as a basketball and a cheap inflatable ball or a whiffle ball and a softball. Push both balls with the same force by hitting them simultaneously with a large book, meterstick, or some other device long enough to hit both balls at once. See which ball travels farther. You all should observe that, if the forces are equal, the object with less mass travels further. (If you have access to a bowling ball, use this as well. It will present a greater contrast between a lighter and heavier ball.)

6. Discuss Newton's third law of motion. Explain that whenever you pull on a thing, it pulls back. In other words, whenever any force is exerted on an object, the object exerts an equal force back. When you pull on a rubber band, you can feel it pulling back. For example, when you catch a baseball, you exert a force on the ball to stop it, and you can definitely feel the force it exerts on your hand. When you swim, you push backwards on the water, and the water pushes you forward.

7. Take the balls outside and let the students kick them (not including the bowling ball, of course) and try to get the balls to roll a specified distance. A larger force will be required for a heavier ball such as a basketball and the students can definitely feel the difference with their feet.

Explanation

An explanation is included in the "Procedure" since this is not a discovery lesson but a directed discussion/demonstration.

Extensions

If you wish to do any of the previous demonstrations as exercises in measurement, you may use the measurement activities as math lessons and perhaps extend the lesson into graphing. All you need to do is have some way of measuring the distance the object moves and some other variable. The other variable could be the mass of the object or the size of the force exerted on it. While you might find it difficult to measure the force exerted directly, you can set up a simple apparatus that will either exert a consistent force or allow you to change the force in a consistent fashion. Here's how: Tie an object to a string and suspend it from a horizontal pole, peg, or the back of a chair. Keeping the string taut, pull the object back and up to a measured, consistent place, then let go. Set the apparatus up in such a way that when the object swings down it strikes another object, setting the second object in motion. (See Figure 1.) You can vary the size of the force by varying the mass of the suspended object or by varying the point to which you pull it back. In either case you need to measure the factor that varies (either position or mass).

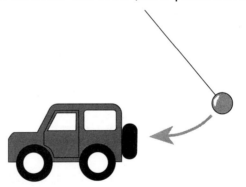

Figure 1: Set an object such as a toy car in motion by striking it with an object suspended from a string.

After you have collected data, you and your students can graph the results. Your horizontal axis would be "Distance Traveled by Struck Object." Your vertical axis might be "Mass of Suspended Object" or "Distance Suspended Object was Pulled Back." As a follow-up activity, you could have a contest to see who could get the car to move and stop at a predetermined point.

Home, safety, and career:
- Discuss the safety implications of moving on any low-friction surface such as a polished floor, ice, or wet, smooth concrete beside a pool. (See the Teaching Science with TOYS activity "Crash Test.")

Language arts:
- Have the students read a book about Newton or Galileo.
- Divide the class into five groups. Have each group read one of the following stories by Seymour Simon. In each story, Einstein Anderson uses his knowledge of motion to solve a puzzle or a problem. The basic story, which is usually about five pages long, is told, then there is a pause in which the reader is supposed to try to figure out what Einstein does. The story continues on the next page with the explanation. After the groups have had time to read their story and do some planning, each group should make a presentation to the class. They might simply tell the story, or they might act it out. After the basic story has been told, the class should brainstorm what Einstein might do to solve the problem. Then the group can reveal what he actually does and why.
  - "The Boat Race" in *Einstein Anderson Lights Up the Sky* (Puffin, ISBN 0-670-29066-1)
  - "The Batty Invention" in *Einstein Anderson Goes to Bat* (Puffin, ISBN 0-14-032303-1)
  - "The Soap Box Derby" in *Einstein Anderson Tells a Comet's Tale* (Puffin, ISBN 0-14-032302-3)
  - "The Weight Lifting Contest" in *Einstein Anderson, Science Sleuth* (Puffin, ISBN 0-14-032098-9)
  - "The Frictionless Roller Skates" in *Einstein Anderson, Science Sleuth* (Puffin, ISBN 0-14-032098-9)

Physical education:
- Discuss topics such as follow-through in kicking or hitting balls in various sports, the need for proper equipment (like appropriate shoes so you won't hold back on a strong kick for fear of what it will do to your foot), or the difference between skating or running any wheeled vehicle on a rough as opposed to a smooth surface.

Further Reading

Bendick, J. *Motion and Gravity;* Franklin Watts: New York, 1972. (Students)

Craig, A.; Rosney, C. *The Usborne Science Encyclopedia;* EDC: Tulsa, OK, 1988. (Students)

Faughn, J.; Turk, J.; Turk, A. *Physical Science;* Saunders College: Philadelphia, 1991; pp 33–44, 193–194. (Teachers)

Kirkpatrick, L.; Wheeler, G. *Physics: A World View;* Saunders College: Philadelphia, 1992; pp 27–45, 464–465. (Teachers)

Handout Master

A master for the following handout is provided:
- Discussion Record

Copy as needed for classroom use.

FORCES AND MOTION

Discussion Record

1. What makes the car move? _____

   _____

   _____

   _____

2. What other ways could you make the car move? _____

   _____

   _____

   _____

3. Name the force causing the motion of each object. _____

   a. A falling ball _____

   b. Paper clips picked up by a magnet _____

   c. Water rocket _____

   d. Block rising in water _____

4. What determines how far the car moves? _____

   _____

   _____

   _____

5. Which ball exerts the biggest force on the kicker's foot? _____

CRASH TEST

Students observe the law of inertia and apply it to automobile safety.

Doll and skate (See the "Extension.")

GRADE LEVELS

Science activity appropriate for grades 4–5
Cross-Curricular Integration intended for grades 4–5

KEY SCIENCE TOPICS

- inertia
- laws of motion

STUDENT BACKGROUND

This activity is designed to introduce the concept of inertia, so students need no previous knowledge of the topic. You may want to do the Teaching Science with TOYS activity "Forces and Motion" first to introduce the basic concept of forces. However, the lesson could also be used as part of a review at the end of a motion unit.

KEY PROCESS SKILLS

- observing — Students observe the movement of a ball or doll on the roller skate as it moves and stops.

- investigating — Students investigate inertia and Newton's first law of motion using a roller skate and ball.

TIME REQUIRED

Setup   negligible
Performance   20 minutes
Cleanup   negligible

Materials

For the "Procedure"
Per class
- roller skate

Old-fashioned, clamp-on metal skates work best, but children's plastic safety skates by Fisher-Price, Tyco, or Larami will work.

- plastic car (large enough for all to see)
- small ball
- (optional) small piece of cardboard
- (optional) small block of wood or other flat-bottomed object

For the "Variation"
Per group of 3–4 students
- shoe box
- small toy car

For the "Extension"
- small doll
- shoestring or piece of ribbon

Safety and Disposal

No special safety or disposal procedures are required.

Getting Ready

1. To save time at the end, you could write the law of inertia on the board and cover it up until the "Procedure," Step 7.

2. Experiment to determine whether the ball rolls better with the skate in its smallest or largest size configuration.

3. For some skates, you may need to cut a small piece of cardboard to lay in the center section to make a more even surface for the ball to roll across.

4. On the plastic skates, cut off the straps or at least fasten them back out of the way.

Introducing the Activity

To arouse interest, pose some questions that you can return to for discussion later. Questions such as the following could be used: "Have you ever wondered why you fall when you stub your toe?" or "Have you ever wondered why you get squashed if you ride in the outer seat in a rapidly spinning amusement-park ride such as the Scrambler?" or "What happens when you are standing in the aisle on a bus and it starts moving?"

Procedure

1. Push along the plastic car (keeping your hand on it) and ask why it is moving. You and the students should establish that it is moving because you are exerting a force on it. Now push the car along and then let it go. Ask why it continued to move even though you stopped exerting a force on it. Discuss.

2. Introduce the generalization that a moving object will keep on moving unless some force causes it to stop. For example, a moving car will keep moving until something stops it. This property of objects is called inertia.

3. If no student points out that the car did stop eventually in Step 1, you should do so. Ask why. Introduce the term friction or remind students of it, and discuss how friction exerts a force. If your class has already done the Teaching Science with TOYS activity "Forces and Motion," remind them how the ball went farther on a smooth (low friction) surface.

4. Ask the students, "What would happen if you could make the surface smoother and smoother, and if you could eventually eliminate friction totally?" The object would keep moving longer and longer. In the absence of friction it would continue forever. Explain that in space there is no friction and so an object, once started, will keep moving indefinitely.

5. Place a small ball in the center of the roller skate. Give the skate a push from the back. The ball will roll toward the back of the skate. Ask the students to describe what they saw. Explain that the skate rolled forward because there was a force on it, but there was no force on the ball. Therefore, although the ball appeared to roll toward the back of the skate, it actually rotated in the spot until the back of the skate caught up with it and pushed it forward. So the skate, which had been set in motion, kept moving, but the ball, which had not been set in motion, didn't move until a force—from the skate—acted upon it.

6. Place the ball at the back edge of the skate and roll the skate toward a wall. When the skate hits the wall, it will stop, but the ball will keep moving. (See Figure 1.) Ask the students to describe what happened in terms of what they've learned about inertia. They (or if need be, you) should note that when the wall exerted a force on the skate and stopped it, that force was not exerted on the ball, so the ball kept on moving.

 You might want to repeat Steps 5 and 6 using a wooden block or other flat-bottomed object in place of the ball. Some students initially believe that because the ball rolls, its behavior is special and that the results would be different for other objects.

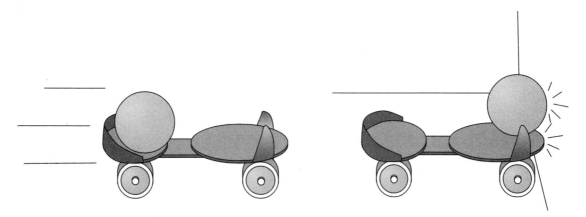

Figure 1: The skate stops, but the ball keeps moving until it too hits the wall.

7. Write Newton's first law of motion on the board or poster and read: An object at rest will remain at rest and an object in motion will remain in motion at the same speed and in the same direction unless a force acts on it to change the motion. Make sure your students understand that the words "at rest" mean sitting still. Explain that the demonstrations illustrate this law.

 If this is a review lesson, you may wish to remind the students of this law at the outset of the lesson.

Explanation

Because of the nature of this lesson, the explanation was included in the "Procedure."

Variation

The Activity Sheet (provided) describes a small group activity that could replace Steps 5 and 6 of the "Procedure."

Extension

To teach a lesson on safety belts in cars, place a doll on the roller skate and roll it into a wall or book. The doll will move to the front of the skate just as the ball did. (See Figure 2.) Explain that this is why you seem to fall forward when a driver slams on the brakes. In case of such a slam, or worse, a collision, you could keep on moving straight out of the windshield if there were not something exerting a force on you to make you stop when the car stops—like a seat belt or an air bag. To demonstrate, tie a shoestring or other "belt" around the doll and have her travel on the roller skate once again. Do this two ways: once with the shoestring just at the waist as a lap belt and once with a lap belt as well as a shoulder harness. Ask the students why the results were different when the doll had "buckled up." Explain that when the doll is tied to the skate, the string exerts a force on the doll and stops the doll from moving separately from the skate—the doll and the skate are forced to act as one unit.

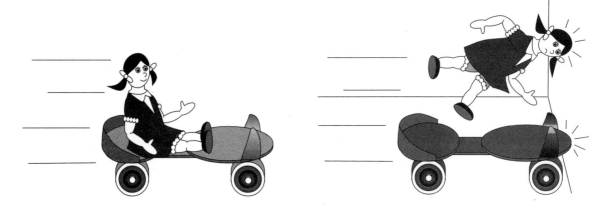

Figure 2: The doll moves to the front of the skate just as the ball did.

Cross-Curricular Integration

Home, safety, and career:
- Have students create a poster or other visual product about seat-belt safety emphasizing the "do's and don'ts." Posters (or other selections) could be displayed in stores, restaurants, and public buildings, particularly in conjunction with National Safety Week or other theme promotions.
- In addition to the seat belt lesson, you can discuss other safety implications. For example, the reason you trip when you catch your foot on something is that your foot is stopped but the rest of you keeps going. Also, the concept of friction helps explain why you will slide on a low-friction surface such as wet smooth concrete next to a pool—and why it is so dangerous to run there.

Language arts:

- After you discuss inertia as defined scientifically, you may wish to discuss a more philosophical definition of inertia. Inertia may be thought of as the tendency of the couch potato to stay on the couch, and the tendency of the active person to stay active. Ask the students it they think people act in accordance with this type of law of inertia. Ask what forces get people to get moving, or to stop. Ask if they can think of anything else that tends to stay the way it is, if no force acts on it. Some ideas are reputations, governments, and habits. You might wish to brainstorm a list of things that act like they are affected by inertia. The students could choose one of these to write about, explaining why they believe—or don't believe—a law of inertia should be written about the item they chose.
- Read and discuss one or more of the following stories by Seymour Simon:
  - "The Broken Window" in *Einstein Anderson Shocks His Friends* (Puffin, ISBN 0-670-29070-x)

    In this story, some students claim the rear bus window was broken by a book that was thrown backwards when the driver hit the brakes. Einstein knows they must be lying, because the book would continue to move forward instead of backward when the bus suddenly stopped.
  - "The Challenge of the Space Station" in *Einstein Anderson Shocks His Friends* (Puffin, ISBN 0-670-29070-x)

    In this story, Margaret is writing a story about the construction of a space station. Einstein points out several scientific errors in her story. One of them is that the construction workers can easily move larger girders around because they are weightless. Einstein knows that they still have a large mass and thus will require a large force to start them moving.
  - "Thinking Power" in *Einstein Anderson Sees Through the Invisible Man* (Puffin, ISBN 0-14-032306-6)

    In this story, Einstein is able to help a friend win a contest because he understands inertia.
- Students can create a TV commercial about seat-belt safety.

Further Reading

Kirkpatrick, L.; Wheeler, G. *Physics: A World View;* Saunders College: 1992. (Teachers)

Hans Jürgen Press. *Simple Science Experiments;* Batsford: London, 1967. (Students)

VanCleave, J. *Physics for Every Kid;* John Wiley & Sons: New York, 1989. (Students)

Contributors

Rainel Dargis, Milford South Elementary, Milford, OH; Teaching Science with TOYS, 1992.

Donna Hooper, Tulip Grove Elementary School, Hermitage, TN; Teaching Science with TOYS, 1994.

Steve Meineke, Milford Main Middle, Milford, OH; Teaching Science with TOYS, 1992.

Richard Willis, Kennebunk High School, Kennebunk, ME; Teaching Science with TOYS, 1994.

Handout Master

A master for the following handout is provided:

- Activity Sheet

Copy as needed for classroom use.

CRASH TEST
Activity Sheet

Each force should be exerted smoothly and quickly. Try not to jerk the box. Repeat each step several times before writing your observations.

1. Place the toy car in the center of the shoe box. Apply a force to the back of the box so that it moves quickly. Record you observations of the motion of the car.

   _____

   _____

2. Place the toy car at the front of the box. Apply a force to the back of the box so it moves quickly. Record your observations of the motion of the car.

   _____

   _____

3. Place the toy car at the back of the box. Apply a force to the back of the box so it moves quickly. Record your observations of the motion of the car.

   _____

   _____

4. Place the toy car at the back of the box. Start the box moving forward, then apply a force to the box to stop it quickly. Record your observations of the motion of the car.

   _____

   _____

YOUR CONCLUSIONS

1. Compare the motion of the toy car for the different positions in the box.

2. When did inertia keep the car at rest? When did inertia keep the car moving?

TWO-DIMENSIONAL MOTION

Students explore motion that occurs in two directions at once.

A yo-yo

GRADE LEVELS

Science activity appropriate for grades 4–8
Cross-Curricular Integration intended for grades 4–6

KEY SCIENCE TOPICS

- forces
- gravity
- inertia
- motion

STUDENT BACKGROUND

This lesson should follow a discussion of the relationship between motion and force and the nature of inertia. You may want to refer to the Teaching Science with TOYS activities "Forces and Motion" and "Crash Test." Students should have some basic understanding of gravity as the force that causes dropped objects to fall.

KEY PROCESS SKILLS

• measuring	Students measure angles and distances.
• controlling variables	Students investigate factors that affect the horizontal distance traveled by a dart.

TIME REQUIRED

Setup	negligible
Performance	50 minutes
Cleanup	negligible

Materials

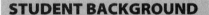

For "Getting Ready" only
- protractor

For the "Procedure"
Part A, per class
- rubber ball

Part B, per class
- yo-yo
- (optional) kickball

Part C, per group of 5–6 students
- soft dart gun or Ping-Pong™ ball gun

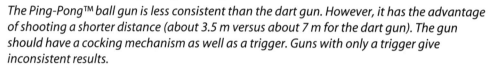

The Ping-Pong™ ball gun is less consistent than the dart gun. However, it has the advantage of shooting a shorter distance (about 3.5 m versus about 7 m for the dart gun). The gun should have a cocking mechanism as well as a trigger. Guns with only a trigger give inconsistent results.

- paper
- measuring tape, meterstick, or several items to use as markers, such as rocks
- black marker
- poster board for large protractor

See construction in "Getting Ready."

For the "Extension"
All materials listed for Part C plus the following:
Per group of 5 students
- small amount of clay

Safety and Disposal

No special safety or disposal procedures are required.

Getting Ready

Prepare protractors:

1. Cut one 12-inch x 12-inch piece of poster board for each group.

2. Starting at the same point each time, draw lines with a black marker at 0, 15, 30, 45, 60, 75, and 90 degrees.

3. Attach a handle made from posterboard to the back of the board.

Procedure

Parts A and B illustrate the relationship between motion and force with some simple demonstrations. Part C provides students with a chance to conduct a motion experiment.

Part A: Simple Direction Changes

1. Review the idea that a force is necessary to make something start or stop moving.

2. Roll a ball into a wall hard enough to make it roll back or bounce off at an angle. Ask the students to describe what they saw.

3. After you establish that the ball changed direction, ask why that happened. Through discussion, establish that the wall exerted a force on the ball. Add to

the rule stated in Step 1 that a force is also necessary to make something change direction.

4. Ask for other examples of a force changing the direction of an object's motion. One example is the baseball, which changes direction when the bat exerts a force on it.

5. Have a student throw a ball directly upward and catch it as it comes back down. Ask the students, "What force changed the ball's direction from up to down?" *Gravity.*

Part B: Motion in a Circle

1. Whirl a yo-yo on the end of a string in a vertical circle. Ask the students if the yo-yo is changing direction. The students may say "no," because the yo-yo is continuing to go in a circle. Explain to the students that a direction is a straight line; a circle is not one direction but a continuous change of direction. (See the geometry lesson in "Cross-Curricular Integration.")

2. Having established that the yo-yo is continuously changing direction, ask, "What force is acting on the yo-yo?" Through discussion, establish that the force is your hand pulling on the string. Mention to the students that the harder you pull, the faster the yo-yo turns. (As you increase the force, the speed increases.) Demonstrate this and have some of the students try it.

3. Explain that the yo-yo was actually undergoing two motions at once. "The yo-yo was going up and down, and at the same time it was moving side-to-side." Ask the students to think of other examples of objects moving in more than one direction at the same time. List their ideas on a board or chart; these may include an airplane taking off, a sled going down a hill, or a ball kicked in kickball or hit in golf.

4. Explain that objects often undergo two motions at once because more than one force acts upon the object (not necessarily at the same time). One force may get the object moving, then another change its direction. For instance, your foot kicks the ball into the air, but gravity causes it to fall toward the ground as it goes forward. Demonstrate this by throwing the rubber ball or kicking a kickball if you have one.

Part C: Dart Gun Experiment

Students should do this experiment outdoors or in the gym. If done outdoors, the day should be windless.

1. Fire the dart gun horizontally. Make sure students note the approximate distance the dart traveled. Ask students to think of something they could do to make the dart go farther. If the students don't suggest tilting the gun upward, do so yourself. Tilt it at roughly 20° above the horizontal and fire. Have the students describe the results.

2. Divide the students into groups of five or six, and give each group a dart gun and a large protractor. Their task is to determine the ideal angle for getting the most distance out of the dart gun.

3. Give the students these instructions:

   a. Keep the gun at the same distance above the ground in each trial. Find a way to make sure the height of the gun is consistent. Some ideas are to use a meterstick, a tabletop, or the same person's belt buckle.

   b. Choose one person to be responsible for watching to see where the dart lands and marking it as soon as possible.

   c. Using the protractor, line up the gun with one of the marked angles and shoot.

   d. Measure the distance the dart traveled, or simply mark the landing spot with a marker. If you use a rock, mark it so you know which rock belongs to which trial.

   e. Repeat Steps c and d three or four times using the same angle. If you are measuring distances, average the results. If you are using rocks to mark the landing spot, estimate the center of the group of rocks.

   f. Repeat Steps c–e with two more angles other than horizontal.

   g. Clearly summarize your observations in a chart or graph (See Figure 1) or by drawing a picture of the floor or ground with your rocks on it, labeling the rocks from each angle. (If a line is desired in the graph, estimate the best-fit straight line or curve as appropriate. The fact that all points do not lie on this line is an indication of experimental uncertainty.)

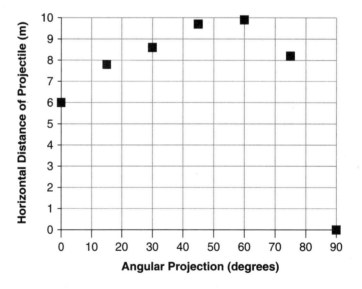

Figure 1: Effect of Initial Angle of Projection on Distance Traveled for a Dart Gun

4. Have each group share their results with the class. Discuss why angles in the middle of the range produce the longest distances.

Explanation

The following explanation is intended for the teacher's information. Modify the explanation for students as required.

What determines how far an object moves in the horizontal direction? The distance is determined by the initial speed of the object, the mass of the object, the amount of friction (including air resistance) the object encounters, the height from which the object is projected, and the angle at which it is projected. All of these factors affect the amount of time the object has to travel before it hits the ground, and thus, the distance it travels horizontally. In the dart gun experiment, four factors are virtually constant each time the gun is shot. The dart gun shoots with the same force, giving the dart the same initial speed; the dart's mass stays constant; there is no variation in the air's friction; and the gun is held at the same height. But the dart does travel different distances when the gun is shot at different angles. To understand why, we need to explore the motion of the dart further.

First, let's think about what happens if we hold the gun horizontally. Once the dart leaves the gun, it continues to travel horizontally at its initial speed. If no other force ever acted on the dart, it would keep traveling forever at the same speed in the same direction. Of course, this doesn't happen. As soon as an object begins moving above the ground, gravity begins pulling it down. (There is also a frictional force exerted on the dart by the air, but this can be ignored for the moment.) Until the object hits the ground, it simultaneously falls and moves horizontally. The distance an object can travel horizontally is determined by how much time it has to travel before it hits the ground. For example, if the dart travels horizontally at 2 feet per second, and it takes 5 seconds to fall to the ground, then it will travel 10 feet horizontally. But if it takes 10 seconds to fall to the ground, it will travel 20 feet horizontally. So if all other factors are constant (force, mass, friction, height), increasing the amount of time the object is in the air will increase the horizontal distance it travels. Objects that are fired, kicked, or thrown upward at an angle take longer to fall to the ground. Thus, they have more time and can travel a greater horizontal distance.

The numerical example above ignores the fact that friction with the air is constantly decreasing the dart's horizontal velocity. It is not necessary to complicate the discussion with air resistance unless you choose to do Variation 2. In this case the subject of air resistance will arise naturally in the process of trying to understand the results.

If throwing a ball at a slight upward angle causes it to travel a greater horizontal distance, what about throwing it at a steep angle or even straight up? If you throw one ball straight up and drop an identical ball at the same time, you will see that the dropped ball hits the floor first. So when you throw something upward, it does take longer to reach the ground. However, you can't maximize horizontal distance by throwing something straight up or at a steep angle. If the angle at which the object is fired, kicked, or thrown is too steep, too much horizontal speed is lost, because most of the force exerted on the object moves it up rather than forward.

So what is the ideal angle for maximizing horizontal distance? The ideal angle will vary somewhat from situation to situation because of differing air resistance effects in each case. The data your students collect shows the best angle in your particular case. Theoretically, 45° is the best angle if there is no air resistance.

An interesting side note to this activity is the idea that the time required for a given object to fall is the same whether it is dropped or fired horizontally. In both cases, the object has no initial vertical velocity. As an aside to Part C, you could have one student drop a dart at the same time another student shoots one horizontally. Both darts should hit the ground at the same time. You can also demonstrate this idea by holding two marbles—or rubber balls or ball bearings—one on top of the other between your thumb and forefinger. Flick the bottom one out horizontally, allowing the other one to fall. You should hear them hit the floor simultaneously.

Variations

1. If you do not have multiple dart guns or Ping-Pong™ ball guns, or you do not have sufficient room for all of your students to work at once, present the lesson in class and then have groups sign up to do the investigation at recess, one per day, until everyone is finished. Then data can be compared and a summarizing discussion held.

2. The dart can also be made to go farther by launching it from a greater distance above the ground. Part of the class could investigate this variable while the rest of the students do Part C. Varying the height from 50 cm to 150 cm works well. Data can be presented in tables or graphs. (See Figure 2.) Your students may find that after a certain point, increasing the height doesn't change the horizontal range. As the dart moves through the air, friction with the air gradually slows it down. Eventually all its forward motion is lost and it falls straight down. After this point is reached, increasing the time it is in the air by raising the firing position does not help it go farther.

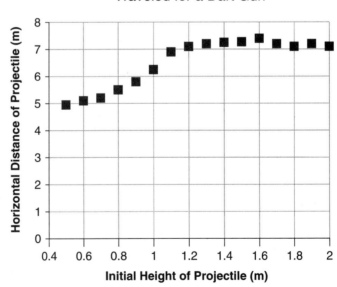

Figure 2: Effect of Initial Height on Distance Traveled for a Dart Gun

Extension

Gradually add mass to the dart (clay works nicely) to see how this affects the motion. (See Figure 3.)

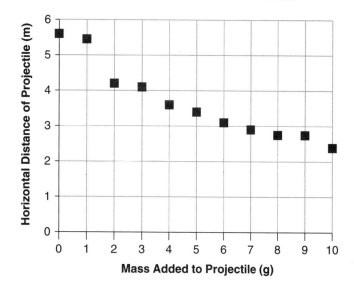

Figure 3: Effect of Projectile's Mass on Distance Traveled for a Dart Gun

Assessment

Options:

- Assess process skills such as controlling variables by watching students collect data.

- Ask students to design an experiment to test a different variable in the dart gun experiment.

- Use the following on a written quiz or as homework. "You and a friend are having a contest to see who can shoot an arrow the farthest. You are both using the same type of bow and arrows. Describe what you can do to make your arrow go farther." Students with poor verbal skills could be given a series of pictures of arrows being shot at various angles and asked to circle the one that would land farthest from the archer.

Cross-Curricular Integration

Language arts:

- Read *Robin Hood* and connect the aiming of the bows and arrows to the two-dimensional motion of the dart guns.
- Have the students write to the local fire department explaining how to get the greatest distance out of their water hoses when fighting a fire (based on what they've learned about two-dimensional motion).
- Have students write stories using the information learned about two-dimensional motion as the solution to the story's conflict. For example, a girl on the softball team is a poor thrower and learns to throw at a 45° angle to become a better player.
- Read and discuss the story "The Science of Baseball Throwing" in *Einstein Anderson, Science Sleuth,* by Seymour Simon (Puffin, ISBN 0-14-032098-9). In this story, Einstein wins a throwing contest by throwing the ball at an angle rather than straight out. Ask the students what is wrong with the sketch showing a ball being thrown horizontally. *It appears to travel in a straight line for a long distance before starting to fall, when in fact it will start to fall immediately.*

Math:

- Part C includes practice in averaging, measuring distance, and reading protractors.
- Discuss the many ways we have of traveling out and back to the same point: We could travel in two directions—forward and back again. We could travel in three directions by traveling a triangular path. If we travel the sides of a square we travel in four directions, and if we travel the sides of a pentagon we travel in five. If your students draw or trace shapes with increasing numbers of sides they can see that as the number of sides increases, the shape more closely approximates a circle. Eventually they may understand that on a circle there is a change in direction each time one moves from one point on the circle to the next.

Further Reading

Faughn, J.; Turk, J.; Turk, A. *Physical Science;* Saunders College: Philadelphia, 1991. (Teachers)

Kirkpatrick, L.; Wheeler, G. *Physics: A World View;* Saunders College: Philadelphia, 1992. (Teachers)

UNDERSTANDING SPEED

Students use the variables of time and distance traveled to determine which toy moves the fastest.

Wind-up walking toys

GRADE LEVELS

Science activity appropriate for grades 4–7
Cross-Curricular Integration intended for grades 4–6

KEY SCIENCE TOPICS

- motion
- speed

STUDENT BACKGROUND

Review use of the meterstick and stopwatch if students have not used them recently. Make sure that students can start, stop, and reset a stopwatch and report time to nearest $1/10$ of a second. Also, have students practice measuring to the nearest $1/10$ cm with a meterstick.

KEY PROCESS SKILLS

• collecting data	Students use stopwatches and metersticks to collect data.
• making graphs	Students graph data in order to determine whether the speed is constant or changing.

TIME REQUIRED

Setup	5	minutes
Performance	90	minutes (two 45-minute periods)
Cleanup	5	minutes

Materials

For the "Procedure"
Part A, per class (when done as a class discussion)
- masking tape
- stopwatch or wall clock with second hand
- meterstick
- several different-sized balls, a variety of toy cars, or wind-up walking toys to race
- stiff cardboard or wood ramps

- books for elevating ramps (unless using wind-up toys)
- several toothpicks
- (optional) ink pad

Part B, per group of 5–6 students
- masking tape
- wind-up walking toy (must walk slowly and in a reasonably straight line)
- stopwatch or wall clock with second hand
- pens to mark masking tape or toothpicks for markers
- 1 or more metersticks or metric measuring tapes

 *Having several metric rulers per group will speed up the data gathering.*

For the "Variation"
Per group of 5–6 students
All materials listed for Part B plus the following:
- several paper clips, dried beans, marbles, or other "units of measurement"
- (optional) pendulum

Safety and Disposal

No special safety or disposal procedures are required.

Getting Ready

Gather objects to race in Part A. These could be different-sized balls, a variety of toy cars, or wind-up toys. Pull-back cars are not suitable here, because they usually move too fast. For best results, if using wind-up toys, test before purchase to determine that they walk in a reasonably straight line. Alternatively, ask students to bring in cars or walking toys.

If the walking toys you have do not walk in sufficiently straight lines for the student to be able to measure how far they walked, you can put down two metersticks side by side to make a path which the toy must walk down. If you are short of metersticks, several layers of masking tape will also make a sufficient barrier to keep the toys walking down the path. Another option is to stamp their feet on an ink pad. The toy will then leave a trail, which can be measured by laying a string along it, then measuring the appropriate length of string.

If you will be working with a large group of students, you may want to enlarge the "Thinking about Motion" Discussion Pictures (provided) for easier viewing. If you are using small groups, make one copy for each group.

Using masking tape, set up a starting line and a finish line on the floor. Set up ramps if you are using balls or toy cars. (If you use wind-ups, you don't need ramps.) You could start the race by pushing the balls or cars, but rolling them down ramps is preferable. Ramps make the motion more repeatable, so that no matter which of the measurement techniques is used the same object is always fastest. Since you want the objects to have different speeds, you can start the various objects at different heights on the ramp or use different angles for the ramps; however, each object should be started from the same place each time.

Introducing the Activity

Show the students the "Thinking about Motion" Discussion Pictures. Ask, "Is the boy moving? How do you know?" *When something moves it changes its position in space. You can see that something is moving by comparing it with other things that are not moving.* Ask a student to move at the rate he or she thinks the boy is moving. Ask several more students to do the same thing (one at a time). Then have the students raise their hands to indicate which they thought was correct. Ask, "What if I told you these pictures were each made an hour apart?" *In that case, he must be moving very, very slowly, so the answer is not quite as simple as it first seemed.* "We must be careful when we describe how fast something moves. For instance, if we compare a child running down the hall to a turtle's motion, then the child runs very fast. But if we compare the child to a car on the highway, he or she runs very slowly. Today we are going to think about measuring and comparing speeds. Let's begin by thinking about how we can tell which object is fastest."

Procedure

Part A can be done either in small groups or as a class discussion with students taking turns helping with the measurements. The "Procedure" is written as if you are using a class discussion. If you use small groups, prepare written instructions that include the questions you want the students to think about as they go along. Part B should definitely be done in small groups. You may want to divide this into two 45-minute class periods rather than one 90-minute lesson.

Part A: Which Toy Is Fastest?

1. Show the students the raceway marked on the floor and the objects that will be the racers. Ask the students, "How can you find out which one goes the fastest?" Someone will probably say, "It is the one that crosses the finish line first."

2. Assign one student to start each object in the race. Assign a couple of other students to watch the finish line.

3. Run the race and determine which racer is fastest.

4. Say to the group, "That was easy, we just had to watch and compare. But how could we determine which was fastest if we raced one at a time?" Hopefully, someone will suggest timing the objects. Run the objects one at a time and measure the time each uses to go from the start to the finish line using a stopwatch or the second hand on your wall clock.

5. Ask the class, "Now that we have these times, how do we determine which is fastest?" If the distances are all the same, then the one with the shortest time is fastest.

6. Ask the students, "Is it possible to determine which is fastest without having them all go the same distance?" After some discussion, measure the distance each object goes in a given time, such as 3 seconds. Do this by having one student watch the second hand or stopwatch and another mark the spot where the object is when "stop" is called. Have the students do this for each object. Now the fastest racer is the one that has gone the greatest distance.

Class Discussion

Ask the students, "What if neither the distances nor the times are the same?"
How you proceed through this discussion will depend on the answers proposed by your students. The following sample discussion illustrates the points you should try to bring out.

If neither the distances nor the times are the same, we have to measure both things to figure out which is fastest—how far in space it moves and how long in time it takes. What if the magnetic car went 4 m in 2 seconds and the roller skate went 9 m in 3 seconds? Which is fastest? How do you decide? Usually we figure out how far each went in the same time period. The magnetic car went 2 m in 1 second and the roller skate went 3 m in 1 second. We say the car traveled at 2 m per second and the roller skate at 3 m per second. This is called its speed. The roller skate had a greater speed than the car. This is what we mean when we say the roller skate is fastest.

Part B: Is the Speed Changing?

1. Introduce this part of the lesson by asking the students, "Do you believe the objects moved at the same speed during the entire motion?" Their answers may not all agree at this point. You may want to introduce the idea that what they have been measuring is the average speed.

In the rest of this activity the students will try to determine if a wind-up toy is walking at constant speed or is speeding up or slowing down.

2. Divide the students into groups of five or six.

Each group will need a Starter, a Timer, and several Measurers to perform the following jobs. Recorder and Reporter could be additional students but these additional jobs can easily be accomplished by the Starter and Timer:
- *Starter—to wind up the toy and let it go;*
- *Measurer—to mark and measure distances;*
- *Timer—to measure times and call out time intervals;*
- *Recorder—to write down all the data gathered; and*
- *Reporter—to report the group's results to the class.*

3. Ask each Starter to wind up the toy and let it go. Have one of the Measurers in each group find the total distance the toy goes before it runs down and each Timer find the total amount of time that elapses before it runs down. (See Figure 1.) Have each Recorder write down the distance and time information.

Figure 1: The Starter winds up the toy and lets it go. The Measurer measures the distance the toy travels before it stops. The Timer times how long the toy takes to run down.

4. Have another Measurer in each group use masking tape to make a path on the floor at least as long as the distance measured in Step 3.

5. Have each Starter wind up the toy and let it go again. Have each Timer call out the time every 5 seconds. At each 5-second interval, have a Measurer from each group mark the position of the toy by marking on the masking tape or laying a toothpick beside the appropriate location. (Since you will have more positions than Measurers, the first Measurer should be prepared to mark again after everyone has marked one position.)

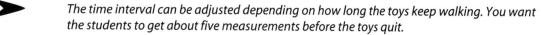 *The time interval can be adjusted depending on how long the toys keep walking. You want the students to get about five measurements before the toys quit.*

6. Have each Measurer find the distance the toy traveled during his or her 5-second interval. Have each Recorder write down this distance and time information.

From this data each group can decide if their toy was moving at a constant speed. If it traveled the same distance in each 5-second interval, then its speed is constant. If it traveled a shorter distance during each successive interval, then it was slowing down.

7. Have each Reporter report the group's results back to the entire class, backing up their conclusions with a data chart or bar graph of distance traveled in each interval.

Explanation

The following explanation is intended for the teacher's information. Modify the explanation for students as required.

As your students discover in this activity, an object's speed tells you how far the object will travel in some standard amount of time, such as 1 second or 1 hour. In order to calculate speed, both distance traveled and time required must be measured. The speed of an object can be constant during an entire motion, or the object may be speeding up or slowing down. If the speed is constant, then the same distance is traveled in every second. If the speed is not constant, the object's average speed can be calculated by dividing the total time interval into the total distance traveled. Average speed for a time interval is that constant speed which would produce the same distance traveled as the actual motion.

Variation

The small group activity in Part B could also be used for a discussion of units. Each group could choose their own units rather than using centimeters. For instance, one group could make a line of paper clips and measure the distance moved in paper clips. Another group might use dried beans or marbles. The units do need to be kept reasonably small. Pencils, for example, would not be a good choice, because students would need to estimate fractions of a pencil. (See Figure 2.)

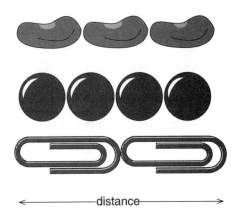

Figure 2: Different items can be used for units of measurement.

If alternative units are used, a Measurer will need to mark the racer's location as the Timer calls out numbers, then go back and count paper clips afterwards. Alternate units of time could also be used. Instead of using a stopwatch and seconds, the Timers could count their heartbeats or the swings of a pendulum. The speed of the toy might then be found to be two paper clips per heartbeat or six marbles per swing. The follow-up discussion should then include the idea that all these different units are equally good as long as the students want to find out only if the toy is slowing down. However, if one group wants to be able to compare their results to another's, then both groups must use the same units. This is why scientists have defined a standard set of units that we call the metric system.

Extensions

1. Students could actually calculate the average speed in each time interval by dividing the distance traveled in each time interval by the number of seconds in each interval. Two sample graphs of speed versus time are provided below. In Figure 3 the speed is almost constant, although it does slow down slightly. The graph in Figure 4 does not indicate a constant speed.

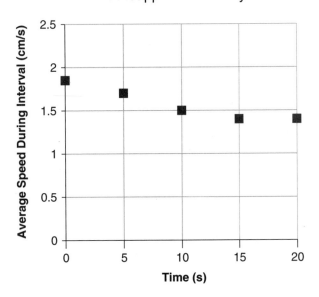

Figure 3: Speed Versus Time Using Pineapple Walker Toy

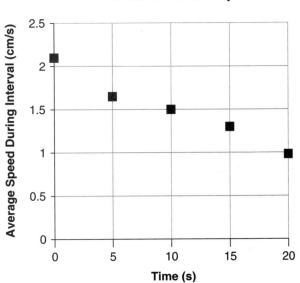

Figure 4: Speed Versus Time Using Dinosaur Walker Toy

2. Have students find the speed of cars driving in front of the school: Measure off a convenient distance, then place a student with a stopwatch at each end. The students should start their stopwatches when the front bumper of the car passes their position. After the car is past, the two students should meet in the middle and stop their stopwatches simultaneously. The difference in the stopwatch readings is the time it took the car to travel the measured distance. This extension might be particularly interesting if done while the school zone speed limit was in force.

Cross-Curricular Integration

Earth science:
- Students can read about the speed at which the crustal plates move.

Language arts:
- Read and discuss something like the tale of the tortoise and the hare or *Fastest and Slowest,* by Anita Ganeri (Aladdin, ISBN 0812062906). Ganeri's book gives examples of record breakers, both fastest and slowest, including animals, plants, trains, and planets.
- Have students write about their recollections of races they have been in or seen.
- Have students write about the activity as if they were the wind-up toys.

Life science:
- Students can compare the speeds at which different organisms move or grow.
- If some parents kept growth records such as marks on a closet door, etc., children could calculate their growth "speed" in inches per year.

Math:
- Measurement and units, proportional reasoning, and averaging skills used in this activity can be connected to math lessons.

Physical education:
- Students can determine the speed of classmates walking, running, and jogging, then determine their heart rates after a certain amount of time in each activity.

Further Reading

Bendick, J. *Motion and Gravity;* Franklin Watts: London, 1972. (Students)

Faughn, J.; Turk, J.; Turk, A. *Physical Science;* Saunders College: Philadelphia, 1991; Chapters 1 and 2. (Teachers)

Kirkpatrick, L.; Wheeler, G. *Physics: A World View;* Saunders College: Philadelphia, 1992; pp 7–12. (Teachers)

Thomas, A.; Langdon, N. *Weighing and Measuring;* Usborne: London, 1986. (Students)

Handout Master

A master for the following handout is provided:
- Thinking about Motion—Discussion Pictures

Copy as needed for classroom use.

UNDERSTANDING SPEED

Thinking about Motion—Discussion Pictures

How fast do you think the boy is moving? Can you be sure?

PUSH-N-GO®

Students are introduced to the concepts of potential energy and energy conversion and apply these concepts to a familiar toy.

Push-n-Go® toy

GRADE LEVELS

Science activity appropriate for grades 4–8
Cross-Curricular Integration intended for grades 4–6

KEY SCIENCE TOPICS

- kinetic energy
- potential energy
- energy conversion
- work
- force
- simple machines
- friction
- laws of motion

STUDENT BACKGROUND

Students should have been introduced to the concepts of work and kinetic energy. They should know that an ordinary stationary object, such as a book, moves in the direction of the force when a force acts on it.

KEY PROCESS SKILLS

- hypothesizing
  Students hypothesize about why the toy does not move in the same direction as the force applied.

- investigating
  Students disassemble the toy to investigate the mechanism.

TIME REQUIRED

Setup	negligible
Performance	30 minutes
Cleanup	5 minutes

Materials

For "Introducing the Activity"
- ball

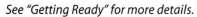

For the "Procedure"

Per small group
- 1 Push-n-Go® toy

See "Getting Ready" for more details.

- 1 screwdriver
- 1 small cup

For "Extensions"

All materials listed for the "Procedure" plus the following:

❶ Per class
- access to several types of horizontal surfaces, such as tile or linoleum, short-nap carpet, long-nap carpet, wood, foam rubber pad, ceiling tile, asphalt, sandpaper, and grass
- meterstick or measuring tape

❷ Per group
- variety of coins or small weights, preferably fairly flat

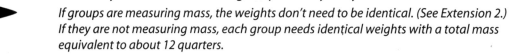

If groups are measuring mass, the weights don't need to be identical. (See Extension 2.) If they are not measuring mass, each group needs identical weights with a total mass equivalent to about 12 quarters.

- tape
- (optional) balance or other device to measure mass

❸ Per class
- other toys in which energy is changed from one kind to another

Safety and Disposal

Depending on the age of your students, you may not wish to have them handle the screwdrivers and take apart the toys themselves. In this case, disassemble several beforehand, but keep one together to demonstrate. No special disposal procedures are required.

Getting Ready

The Push-n-Go® by Tomy America is available at many toy stores, is inexpensive, and comes in a variety of shapes that have basically identical internal mechanisms. The styles include fire engine, dump truck, airplane, dinosaur, and others. It is marketed as a preschool toy, so many of your students may remember having one when they were younger. Some of them may be able to bring one from home for a few days. A number of other companies make toys with similar inner mechanisms. Press-N-Go™ Fire Engine by Shelcore®, Press-N-Go Cookie Monster by Illco® and Go-Go Gears™ by Playskool® are examples. The Push-n-Go® comes apart somewhat more easily, but the others could also be used in this lesson.

You may want to loosen all the screws and then retighten them to make sure that none are stuck. This is more important with younger children. Collect or identify the surfaces if you are planning to perform Extension 1.

Introducing the Activity

Gently roll a ball away from you. Have a student roll it back. Listen to students' observations and help them to conclude that you pushed away, and the ball moved away. Review the rule that when a force acts on a simple stationary object, the object moves in the direction of the force.

Procedure

1. Demonstrate the Push-n-Go® by pushing down on the rider's head and releasing it. Ask a student to demonstrate it a second time. Ask how this is different from the ball demonstration you just did. Listen to students' observations and help them to conclude that this time the force was downward and the movement was forward. Establish that, in this case, the object did not move in the direction of the force.

2. Ask the students to hypothesize about why the toy did not move in the direction of the force. Listen and restate, helping them to clarify their ideas. Guide the group to develop the hypothesis that something within the toy stored the energy and then pushed in a different direction. Ask the students to suggest what might be in the toy that stored the energy and changed the direction. Have them record suggestions.

3. Tell the students that they are going to take the toys apart to check out their hypotheses.

    You may want to tell them about the homework assignment at this time. See Step 7.

   Divide the class into groups. Direct each group to use the screwdriver to remove the screws from the bottom of the Push-n-Go®. Have each group put the screws in their small cup. They should all be able to see the inside of the toy well.

4. Have a student in each group push the rider down on the spring. Tell the groups to watch what happens, discuss what is occurring, and write a description of these events in their own words. Make sure they explain both how the energy is stored and how the direction is changed. Encourage students who see and clearly understand what is happening to explain it to those who may be looking but not understanding.

5. Ask the students to discover how the gears are prevented from turning while the spring is being pushed down. Ask why this is necessary.

6. Through discussion, bring the class to consensus on how the toy stores energy, changes the direction of the force, and keeps the gears from turning while the head is being push down. (See the "Explanation.")

7. Have the students reassemble the toys, run them again, and practice narrating what is happening inside the toy as they use it. Explain to the students that their homework assignment is to share the toy with someone at home or a friend and explain how it works.

8. Set up a schedule for each student who does not own a Push-n-Go® to take the toy home. If desired, give each student a form to take home for the signature of a friend or family member who has listened to the explanation and observed the toy.

   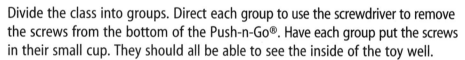 *Your students will learn a lot from practicing, focusing on, and delivering the explanation.*

Explanation

 The following explanation is intended for the teacher's information. Modify the explanation for students as required.

When you operate the Push-n-Go®, you apply a downward force to the rider's head and it moves downward as a result. Thus, an applied force has moved through a distance and work has been done. As you push, the head pushes back on your finger, showing that a force is being exerted. Clearly, the movement of the head downward occurs while the force is being applied.

Underneath the rider is a rather stiff spring. As the rider moves down, the spring compresses, storing energy. Once the head goes as far down as possible, nothing happens until you remove your finger.

 If you are demonstrating and explaining at the same time, this is a good time to hold your finger still and mention that your work was converted into potential energy that is now stored in the spring. Your body's energy was transferred, through the process of doing work, to the spring.

When the rider is released, the spring extends, and it pushes the rider back up. As the rider rises, a rack (series of teeth or notches) on the rider exerts a force to turn a pinion (gear) that is mounted on the axle. (See Figure 1.) The pinion transfers this force through all the other gears to the wheels causing the Push-n-Go® to roll forward. You can turn the axle by hand to see the gears move in slow motion. The removal of one additional screw on the side of the gear housing detaches it from the base of the truck, making the lower two gears more easily visible.

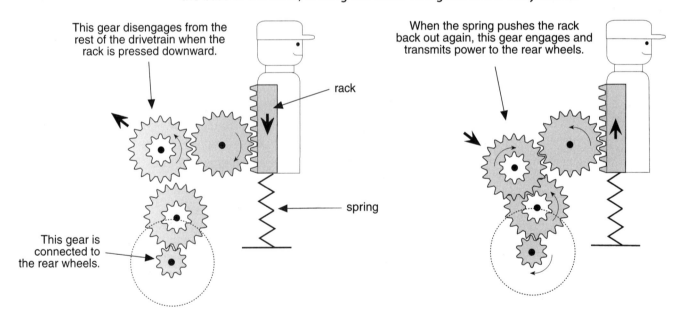

Figure 1: When the rider is pushed down, the spring compresses, storing energy. When the rider rises, a rack on the rider causes the gears to turn.

The process can be described in energy terms as follows: the potential energy of the spring is converted into kinetic (or motion) energy of the gears and the truck as a whole. A small part of the energy is converted to heat energy by friction between the parts.

Since the rack causes the axle to turn while the rider is being pushed up, why doesn't the same thing happen when the rider is being pushed down? The motion of the axle while the spring is being compressed is prevented by an interesting feature of the middle gear. This gear is mounted in a slot that allows about half a centimeter of vertical travel. When the spring is extending, the pinion pushes down on the middle gear, causing it to turn and to make contact with the gear below it, but when the spring is being compressed, the pinion turns in the opposite direction and lifts up on the middle gear, causing it to move forward. Thus, no contact is made between the middle gear and the gear below, and the axle does not turn.

When the Push-n-Go® stops rolling, you might think the energy has disappeared. However, the energy is not lost. Rather, it has changed into heat energy because of the friction between the wheels and the floor. On a carpeted floor where there is a greater amount of friction, the Push-n-Go® will not travel as far before all the kinetic energy has been changed to heat. If two students send the Push-n-Go® rapidly back and forth, the rubber rings on the wheels will begin to feel warm. This observation helps students verify that the kinetic energy is not lost but has been changed into heat energy.

Extensions

1. Surfaces Extension

a. Remind the students that the friction between the toy and the floor caused the toy to stop running after it had gone about 5 m. The kinetic energy was changed into heat energy. Ask them, "What would happen if the surface had more friction? What if it had less?"

b. If your students have had sufficient experience, you can ask them to design and carry out an experiment to determine which of several surfaces exerts the largest friction force on the toy. If they have not had sufficient experience, follow Steps c–e to lead them through the design and testing process step by step.

c. Explain to the students that they will be comparing the way the Push-n-Go® runs on several different surfaces (that have differing amounts of friction) by measuring the distances traveled by the toy on each surface.

d. Ask the students to choose several surfaces from those that you have available to use in the experiment.

e. Remind the students that a good experiment has only one variable; all other factors that could affect the results are held constant. Ask students, "What is the variable in this experiment?" *The variable is the surface.* Explain that since some Push-n-Gos® may run better or worse than others, the same Push-n-Go® should be used for all the work by each group.

Tell students that another feature of a good experiment is a hypothesis. Have each group write a hypothesis on the line at the top of the "Surfaces Extension" Data Sheet (provided). This hypothesis should state which surface they think has the least amount of friction, and therefore on which surface the Push-n-Go® will go the farthest.

f. Review the format for the Data Sheet with students.

g. Have each group run their toys several times on each of the surfaces they have chosen to test. Have students record the distances traveled on the chart, compute averages, and plot a bar graph. (See Figure 2.)

Figure 2: Distance Traveled by

Push-n-Go® Toy on Different Surfaces

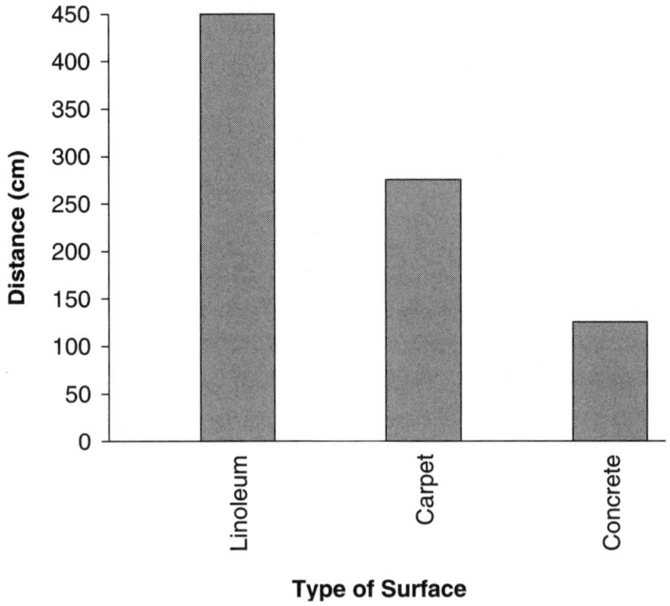

h. Have each group list the surfaces they tested in order of longest roll to shortest roll. Ask the students, "Do all results agree? If not, what might have caused the difference?" Discuss the results in terms of the size of the friction force exerted by each surface.

2. Weights Extension

a. Before doing this extension, your students should know that if a push is constant, varying the weight of an object is one way to vary its distance traveled. This extension will demonstrate this idea using the Push-n-Go®. You could introduce the extension by pushing a book, ball, or toy car and asking, "How might we vary the distance an object goes?" Students are likely to answer that pushing it harder or more gently would vary the distance. "Pushing it harder—or more gently—does not work for the Push-n-Go®. What would work?" Through class discussion, the idea should emerge that varying the mass of the Push-n-Go® will alter the distance it goes.

b. Remind your students that in a well-designed experiment, we examine only one variable at a time. Ask students, "What is the variable in this experiment?" *The variable is the mass of the toy.* Students should remember to keep everything else constant, such as the specific Push-n-Go® used and the surface on which it runs.

Have each group of students write a hypothesis at the top of the Weights Extension Data Sheet (provided). A simple hypothesis such as, "The more massive toy won't go as far," is fine. If students wish to hypothesize about the actual distance change for a certain amount of extra mass (for example, 1 foot less if carrying four quarters), encourage them to do so.

Teaching Physics with TOYS

c. Have each group first establish a baseline distance for the regular Push-n-Go® by finding its mass, then running it three times, measuring the distance, and averaging. Have students record the results on the "Weights Extension" Data Sheet (provided).

Determining the mass is not absolutely necessary. If you use identical objects such as quarters in Step d, then you could just count the number of quarters.

d. Have each group load some weights, coins, or small rocks into the toy or tape them to the outside of the toy. Have each group find the mass of the entire object now and record the results. Each group should run it three times, measuring the distance, and record the results on the "Weights Extension" Data Sheet.

e. Have each group repeat Step d using a different amount of weights, coins, or rocks than before.

f. Have the students graph the results on a line graph, with the mass (or number of quarters) as the horizontal (x) axis and distance traveled as the vertical (y) axis. The graph is started on the data sheet. (See Figure 3.)

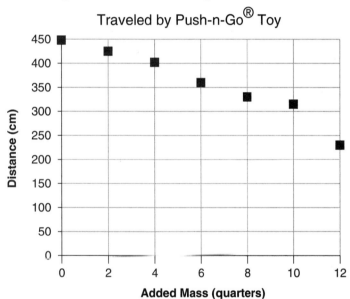

Figure 3: Effect of Adding Mass on Distance Traveled by Push-n-Go® Toy

3. Energy Conversions Extension

Ask students to give examples of other toys in which energy is changed from one kind to another. You might ask the students to bring in toys the next day that illustrate energy conversion. Many toys fit this criteria. All toys using springs, whether compressed or wound, could be analyzed in a manner similar to the Push-n-Go®, as could even simpler toys such as paddleballs and bouncing balls. Any toy that uses batteries converts electrical energy into kinetic energy, light energy, or sound energy.

Assessment

Options:

- Informally observe students during performance of the activity and during class discussion and formal evaluation of data collection and graphs.

- For an activity-based evaluation, students could examine the workings of the manual pencil sharpener in the classroom and describe how muscle energy is converted such that the result is a sharpened pencil. Be sure to instruct students to include in their explanations the workings of the pencil sharpener's gears and grinders.

Cross-Curricular Integration

Language arts:
- Have students imagine the following scenario: Push-n-Go® toys have been popular for a number of years because of their unique design and durability, but recently the sales have slumped. You have the following assignment as an advertising agent: Describe your designs for the new and improved Push-n-Go® toy line. Include in your discussion: 1) why you think this design will sell; 2) who your targeted buyers are; and 3) a possible price for the product. Finally, make up an advertisement for the new design and include illustrations.

Math:
- Students use measurement, graphing, and charting skills in the Extensions.

Further Reading

Faughn, J.; Turk, J.; Turk, A. *Physical Science;* Saunders College: Philadelphia, 1991. (Teachers)

Gartrell, J.E.; Schafer, L.E. *Evidence of Energy;* National Science Teachers Association: Washington, D.C., 1990. (Teachers)

Kirkpatrick, L.; Wheeler, G. *Physics: A World View;* Saunders College: Philadelphia, 1992. (Teachers)

Handout Masters

Masters for the following handouts are provided:
- Surfaces Extension—Data Sheet
- Weights Extension—Data Sheet

Copy as needed for classroom use.

Name _____   Date _____

PUSH-N-GO®

Surfaces Extension—Data Sheet

Hypothesis:

_____

_____

_____

CHART

Trial Number	Distance Traveled			
	Surface One: _____ (Write name of surface.)	Surface Two: _____ (Write name of surface.)	Surface Three: _____ (Write name of surface.)	Surface Four: _____ (Write name of surface.)
1				
2				
3				
Average				

GRAPH

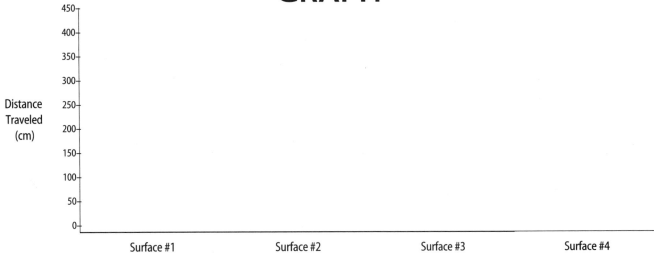

Name _____   Date _____

PUSH-N-GO®

Weights Extension—Data Sheet

Hypothesis:

_____

_____

_____

CHART

Trial Number	Distance Traveled				
	No extra load _____ (Mass)	Extra load 1 _____ (Mass)	Extra load 2 _____ (Mass)	Extra load 3 _____ (Mass)	Extra load 4 _____ (Mass)
1					
2					
3					
Average					

GRAPH

Distance Traveled (cm)

450 — 400 — 350 — 300 — 250 — 200 — 150 — 100 — 50 — 0

Total Mass or Number of Quarters

THE TOY THAT RETURNS

Explore the concepts of elastic potential energy and kinetic energy using rubber bands as "storers" of energy.

A commercial come-back toy

GRADE LEVELS

Science activity appropriate for grades 4–8
Cross-Curricular Integration intended for grades 4–6

KEY SCIENCE TOPICS

- elastic potential energy
- kinetic energy
- motion
- work
- inertia

STUDENT BACKGROUND

Students should have already been introduced to the concept of kinetic energy and elastic potential (or stored) energy. This activity is a good follow-up to the Teaching Science with TOYS activity "Push-n-Go®."

KEY PROCESS SKILLS

• inferring	Students infer that a commercial come-back toy works because of a stored energy system.
• investigating	Students investigate the mechanism that powers the come-back toy and build their own.

TIME REQUIRED

Setup	15 minutes
Performance	40 minutes
Cleanup	5 minutes

Materials

For the "Procedure"
Per student (ideally) or per group of 4–5 students (plus 1 set for teacher)
- small (1 pound) coffee can with both metal ends removed

Larger cans can be used but will require larger rubber bands. Have students start collecting cans and plastic lids several weeks in advance.

- 2 plastic lids for the coffee can
- rubber band about 8–10 cm long
- scissors
- 1 piece of wire or twist tie about 10 cm long
- a weight such as 1 of the following:
  - bolt and several nuts
  - larger teardrop fishing sinker
  - pennies tied up in a piece of cloth
- 2 toothpicks

Per class

- (optional) 1 or 2 windup toys with visible rubber bands

Windup airplanes often have visible rubber bands. (See the Teaching Science with TOYS activity "Delta Dart.")

- commercial come-back toy such as Come Back Roly Toy by Shackman (See photo), Mickey Mouse Roll Back Wheel by Illco, or Rollback Pals™ by Child Dimension™

For "Variations and Extensions"
❶ Per class
- (optional) 1 bell, buzzer, or whistle

❷ Per group
- masking tape
- meterstick
- coffee can

Safety and Disposal

Cover sharp edges on the cans with electrical or masking tape. No special disposal procedures are required.

Getting Ready

1. Assemble one come-back toy as an example of the finished product. (See Figures 1–4.)

   a. Cut out both metal ends of small coffee can.

   b. Use scissors to punch a hole in the center of each of the plastic lids.

   c. Feed rubber band ends through each hole from the inside of the lid and insert a toothpick through the loop of the rubber band to keep it from pulling back through. Stretch the rubber band to pull the toothpicks up snugly against the lids.

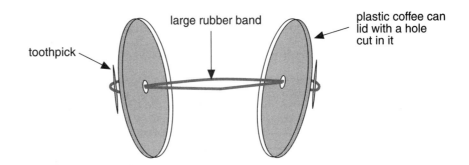

Figure 1: Cut a hole in each plastic coffee can lid and string a rubber band between them.

d. Tie the wire or twist tie to the weight.

e. Tie the wire or twist tie to the lower section of the rubber band. (See Figure 2.) The wire should be short enough so that the weight will not touch the side of the can.

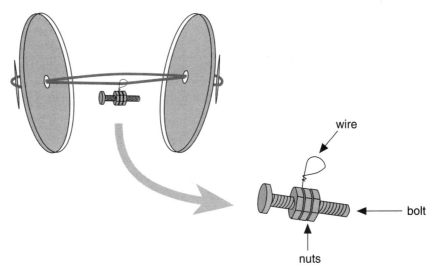

Figure 2: Construct the weight and attach it to the rubber band.

f. Feed one plastic lid through the coffee can (See Figure 3) and snap it over the rim of the can. Snap the second lid over the rim on the other end. The toy is complete.

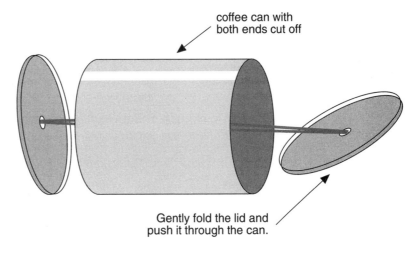

Figure 3: Push one of the lids through the coffee can.

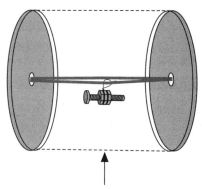

Make sure that the rubber band is tight enough to keep the weight from touching the side of the can.

Figure 4: Snap both lids onto the can.

   g. Try out the toy: Place it on its side on a flat surface and give it a push to start it rolling. If it does not return, minor adjustment may be needed. Make sure that the weight isn't touching the side of the can. Try adding more weight or tightening the rubber band.

2. (optional) With younger students, you may wish to punch holes in all the can lids in advance.

Procedure

1. Demonstrate the motion of the commercial come-back toy by giving it a push. Ask some questions to review the concepts involved, such as, "What kind of energy does the toy have while it is moving? Where does that energy come from? Where does the energy go when the toy slows down and stops? Why does the toy return?"

2. Lead the students to speculate about what is inside the toy that stores energy. If no one mentions rubber bands, demonstrate another stored-energy toy in which the rubber band is visible.

3. After some ideas about what's storing the energy have been proposed, ask the students how they could find out which one is correct. Hopefully someone will recommend taking the toy apart. If your commercial come-back toy can be taken apart, open up the toy and let the students come up in small groups to see it or pass it around.

4. Have the students construct homemade come-back toys either individually or in small groups by following the "Make a Come-Back Toy" Instruction Sheet (provided).

Troubleshooting toys that don't work is an important part of the students' learning process. It helps them to understand what features in the design are essential to making the toy work and is a good problem-solving exercise.

5. Have the students write explanations of how the come-back cans work.

Explanation

The following explanation is intended for the teacher's information. Modify the explanation for students as required.

When you do work by pushing the can, you give the can kinetic energy. Because the weight does not turn with the can, the rubber band winds up. Rubber bands store energy when they are stretched or twisted and release it later. The energy stored in the twisted rubber band is called elastic potential energy. This potential energy is released and returns to kinetic energy as the rubber band unwinds, rolling the can back to you. The can continues to roll back even after the rubber band is completely unwound. This is due to the can's inertia. This motion winds the rubber band in the opposite direction, again storing elastic potential energy, stopping the can again, and rolling it forward. This process continues until friction converts all the kinetic energy into thermal energy, and the can stops.

Variation

A commercial come-back toy (with a hidden mechanism) is a good tool to use for convergent thinking. To use it in this way do the following: 1) Demonstrate the toy several times. Pet it and/or whisper to it to add a little drama. You can also whistle to call it back. 2) Have the students ask simple "yes" or "no" questions, one at a time, about the toy's construction. 3) Use a bell or buzzer to indicate questions that correctly describe the construction of the toy. Do not respond to questions to which the answer is no, or complex questions to which the answer would be partly yes and partly no. Students must LISTEN and BUILD on each others' questions to reach a solution. 4) When a student believes that he/she can describe the construction from beginning to end, give that student an opportunity to share his/her hypothesis with the class.

Extension

Put a strip of masking tape on the floor. Mark the starting point near the middle. As the can rolls back and forth mark each point at which it turns around. Measure from the starting point to each turning point. Graph these distances taking right of start to be positive and left negative or vice versa. (See Figure 5.)

Figure 5: Turn-Around Distances

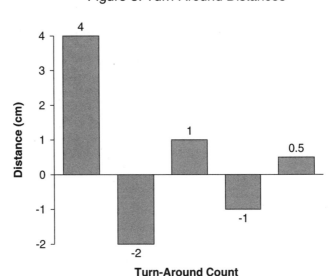

Cross-Curricular Integration

Language arts:
- Have students list the many varied and unusual uses for a rubber band.
- Tell students, "You are a rubber band. Write a story of your life. Be sure to include your thoughts, feelings, and reactions. Begin at birth and continue through life."
- Have students do "I Search" papers. Tell students, "Think of a question about rubber bands or rubber that you would like answered. Research your questions using unconventional methods." Some examples are personal interviews, writing letters, and phone calls.

Life science:
- Discuss the way the body stores chemical energy and uses it later to move muscles.

Math:
- Have students measure the distance the toy rolled forward and then backward. Find the ratio of the two distances. Ask students, "Is this ratio always the same?" This could provide practice in long division, fractions, or percentages.

Further Reading

Allison, L.; Katz, D. *Gee Wiz! How to Mix Art and Science or the Art of Thinking Scientifically;* Little, Brown: Boston, 1983. (Students)

Faughn, J.; Turk, J.; Turk, A. *Physical Science;* Saunders College: Philadelphia, 1991, pp 59–66. (Teachers)

Kirkpatrick, L.; Wheeler, G. *Physics: A World View;* Saunders College: Philadelphia, 1992, pp 117–132. (Teachers)

Contributor

Mark Beck, Indian Meadows Primary School, Ft. Wayne, IN; Teaching Science with TOYS peer mentor.

Handout Master

A master for the following handout is provided:
- Make a Come-Back Toy—Instruction Sheet

Copy as needed for classroom use.

THE TOY THAT RETURNS

Make a Come-Back Toy—Instruction Sheet

1. Thread the rubber band through the holes in the lids, and secure with toothpicks.

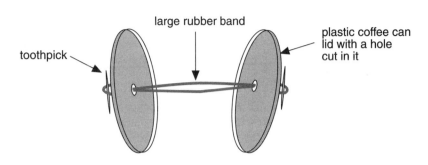

toothpick

large rubber band

plastic coffee can lid with a hole cut in it

2. Tie the weight on the rubber band.

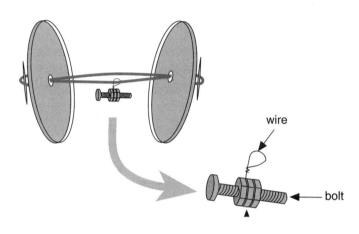

wire

bolt

3. Pull one lid through the can and snap on both lids.

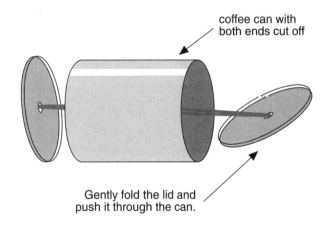

coffee can with both ends cut off

Gently fold the lid and push it through the can.

4. Try rolling your can. If it does not come back, adjust the weight or tighten the rubber band.

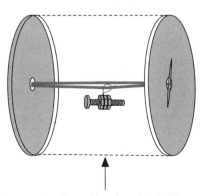

Make sure that the rubber band is tight enough to keep the weight from touching the side of the can.

PHYSICS WITH A DARDA® COASTER

Students use Darda® coasters to explore gravitational potential energy, kinetic energy, energy transformations, and centripetal force.

A Darda® car on a loop-the-loop track

GRADE LEVELS

Science activity appropriate for grades 5–8
Cross-Curricular Integration intended for grades 5–6

KEY SCIENCE TOPICS

- gravitational potential energy
- kinetic energy
- centripetal force

STUDENT BACKGROUND

Students should be familiar with kinetic energy, gravitational potential energy, and the idea that energy can be transformed from one type to another. If students have previously studied circular motion, the concept of centripetal force can be reviewed and reinforced with this lesson. If not, that section may be omitted.

KEY PROCESS SKILLS

- observing — Students carefully observe a Darda® car on a loop-the-loop track.

- predicting — Students predict what factors will allow the car to successfully complete the loop.

- hypothesizing — Students form hypotheses as to why the car sometimes fails to "make the loop."

TIME REQUIRED

Setup	10	minutes
Performance	20	minutes
Cleanup	5	minutes

Materials

For the "Procedure"
Per class
- pull-back car
- loop-the-loop track

 These are manufactured by Darda®, Majorette®, Hot Wheels®, and others.

For "Extensions"
❶ Per class
- loop-the-loop track with jump
- pull-back car

❷ All materials listed for the "Procedure" plus the following:
Per class
- measuring device, such as meterstick or tape measure

Safety and Disposal

No special safety or disposal procedures are required.

Getting Ready

1. Assemble the track.

2. Prepare a handout on which the students can respond to questions concerning the experiment. (A sample Observation Sheet is provided.)

3. If doing the "Variation" or "Extensions," prepare index cards. (A template for Instructions for Small Group or Learning Center Investigation is provided.)

Introducing the Activity

Demonstrate the coaster several times. Vary the car's speed so that sometimes it goes around the loop and other times it does not. You may wish to appoint a student to catch the car at the end of the end of the track.

Procedure

1. Ask students, "What determines whether the car goes around the loop?" *Initial speed, initial kinetic energy, and how far you pulled it back (or amount of energy initially stored).*

2. Ask students, "Why does the car slow down as it goes up the loop?" *It is gaining gravitational potential energy, so it must lose kinetic energy or energy of motion.*

3. Ask students if there is any way to make the car go around the loop without winding up its motor. If no one comes up with a suggestion, ask them to think about a real roller coaster. Ask, "How does it work?" *At the beginning of the ride, the cars are pulled up a high hill, then they gain speed as they go down it.* Lift up the end of the track to make a tall ramp. Determine how high up on the track you must place the car in order for it to make the loop.

 This time you are storing energy in the car using gravity rather than winding the motor.

4. Demonstrate several unsuccessful loops using either method. Ask the students, "Why did the car fall?" Make sure they notice that the car does not stop before falling, so the problem is not that it loses all its energy of motion and falls. Make the connection to the "weightlessness" of the astronauts in the space shuttle. If the car is going fast enough that the centripetal force required is greater than the weight of the car, then it makes the loop.

 Step 4 can be omitted if the students do not have previous experience with centripetal force.

Class Discussion

Ask the students to identify when the car

1. has the most kinetic energy,

2. has the most gravitational potential energy,

 This answer differs depending on which way you start the car.

3. is speeding up,

4. is slowing down, and

5. experiences the largest centripetal force. (Include this only if you do Step 4 of the "Procedure.")

Explanation

The following explanation is intended for the teacher's information. Modify the explanation for students as required.

In order to make the car move, you must give it energy. You can do this by pushing it, winding its motor, or lifting it (and the track) up. As the car begins to go up the loop, it gains gravitational potential energy and, thus, must lose kinetic energy (energy of motion). If the car makes it around the loop, it gains this kinetic energy back on the way down, except for a small amount used up as work done to overcome friction. If it does not make it around, it still has some kinetic energy left when it falls. It does not fall because it lost all its energy.

So why does the car sometimes fall? To understand why, we need to look at the two forces acting on the car as it goes around the loop—gravity and the force of the track against the wheels of the car. Together, these forces provide the centripetal force that keeps the car moving in a circle.

Just before the car enters the loop, it is moving along a straight path. If no new force was exerted on the car, it would continue along this straight path forever. For the car to change direction and start moving in a circle, a force must act on it. As the car enters the loop, a force is exerted against its wheels by the track. (See Figure 1.)

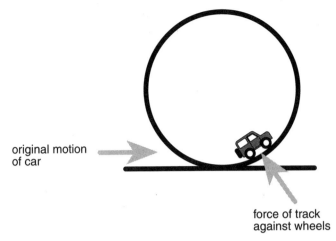

original motion
of car

force of track
against wheels

Figure 1: The track exerts a force that changes
the direction of the car's motion.

The speed and mass of the car and the diameter of the loop determine the size of
the force the track must exert to make the car change direction and start moving
in a circle. The faster the car is moving, the greater the force of the track must be.
As the car continues to move around the loop, it is continually changing direction
and requires a continuous force towards the center of the circle (centripetal force)
to make it do so.

You can visualize the role of centripetal force by imagining what would happen if,
at any point in the car's path around the circle, a car-sized hole suddenly appeared
in the track, right in front of the car. The car would shoot out of the hole, moving
in a straight line tangent to the circle. (See Figure 2.) If there is no force acting on
the car, it continues to move at the same speed and in the same direction.

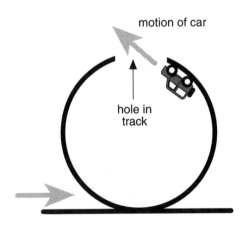

motion of car

hole in
track

Figure 2: If a hole appeared in the track, the car
would travel in a straight line out of the loop.

As the car goes around the track, two changes in its motion take place. As we just
saw, the car is changing direction. It is also slowing down as some of its kinetic
energy is being converted into gravitational potential energy. Both these changes
must be produced by a force. The direction change requires a force towards the
center of the circle (perpendicular to the track) and the speed change requires a
force directed opposite the direction of the motion (tangent to the track). Where
do these forces come from? The force perpendicular to the track comes from two

sources: One component is the force the track exerts on the car, which is at all points perpendicular to the track (just as the force a table exerts on a book which is laying on it is perpendicular to the table). Another component of the perpendicular force comes from the force of gravity. The force of gravity always acts vertically down on the car. To understand the effect of the gravity force, we need to think of it as consisting of two parts, one perpendicular to the track (p) and one tangent to the track (t), which add together to make the actual downward force (g). (See Figure 3.) Thus, the force of gravity also provides the force tangent to the track.

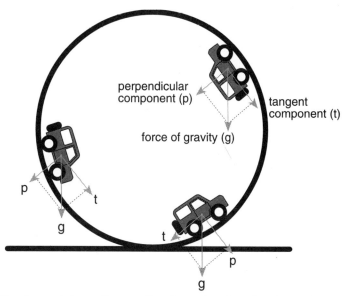

Figure 3: The force of gravity can be thought of as having two parts.

The relative size of the perpendicular and tangent parts of the gravity force changes as the car moves around the circle. (See Figure 3.) Near the bottom or top of the track the tangent part is small and the perpendicular part is large. Near the middle of the track, the tangent part is large and the perpendicular part is small. As mentioned previously, the part of the gravity force tangent to the track is the force that changes the speed of the car. The part of the gravity force perpendicular to the track is added to the perpendicular force of the track pushing on the car and together they provide the force needed to make the car change direction. Any force that makes something move in a circle is called a centripetal force, so this combination of perpendicular forces is often referred to as the centripetal force even though it is not a single force.

We have ignored friction between the track and the car, because it is small and because it would not significantly change the discussion.

Remember that for the car to move in a circle, the net perpendicular force must be inward. When the car is in the bottom half of the loop, the part of the gravity force that is perpendicular to the track points away from the center of the circle. For the net perpendicular force to be inward, the track must exert a greater force than the perpendicular portion of the gravity force. When the car is in the top half of the loop, the perpendicular gravity force points toward the center of the circle. As a result, the track force can be smaller and still result in a net perpendicular force inward.

Now we are ready to answer the questions with which we began: Why does the car sometimes fall before it goes all the way around the loop? The faster the car goes, the larger the size of the centripetal force needed to change the direction of the car's motion. If the car has a high speed, then the centripetal force needed is always larger than the perpendicular part of the gravity force and the track pushes inward to provide the additional force. Thus, the car stays in the loop. If the car is moving slowly, then the centripetal force needed may be smaller than the perpendicular part of the gravity force. The track does not need to push inward. In fact, part of the gravity force is "left over;" it isn't needed to move the car in a circle. The track can't grab the car and pull up to cancel the extra gravity force, so the car falls.

In Extension 1 you jump the car over a gap in the track. This activity involves two-dimensional motion. As the car approaches the jump, it is moving horizontally. As soon as the car leaves the track, it continues to move horizontally and also begins to fall. Because the car is falling, the track on the landing side must be lower than the track on the first side (if the track leading up to the jump is level). Whether a car makes a particular jump depends on the size of the gap and the car's initial horizontal speed. If the car crosses the gap before it has fallen below the level of the track on the landing side, then it lands safely. If the car falls too far before reaching the landing side, it crashes into the gap.

If the track leading up to the jump is slanted slightly upward, the car leaves the track moving upward as well as horizontally. As the car moves horizontally, it will continue its upward motion, gradually slowing, and eventually beginning to fall. Because the car moves upward initially as it crosses the gap, it may not be below the starting height when it reaches the other side. So the track on the landing side may not have to be lower. Also, the upward motion enables the car to stay in the air longer, so the gap can be wider.

Variation

If you prefer a hands-on learning center to the demonstration described in the "Procedure," give the students the task of figuring out two different ways to make the car go around the loop. Follow this with any or all of the questions from the "Class Discussion" as appropriate. (Instructions for a small group or learning center investigation are provided in case you want a longer, more structured assignment.)

Extensions

1. Some of the brands of coasters have jumps in them. Discuss why the car is able to cross the jump. Students could also investigate what happens if you make the jump wider, what happens if one side is higher than the other, and if it matters whether the track on either side of the jump is level rather than slanted. (Student instructions for a hands-on learning center are provided.)

2. The pull-back cars can be used for a hands-on activity involving measurements if students pull the cars back set distances to wind them and then determine (a) if the car makes it through the loop and, (b) if the car does make the loop, how far it travels afterwards. One interesting question to ask is, "Do you reach a

point where further winding does not make any difference?" Also, "Is there a linear relationship between winding distance and travel?" (That is, "Does pulling it back twice as far make it go twice as far?") This part could also be done without the track by running the car on the floor. (Instructions for small group or learning center investigation are provided.)

Cross-Curricular Integration

Language arts/Art:
- Having used the pull-back car in a number of ways by this time, students could make advertisements for the toy car and the track as a language arts/graphic arts/drama activity. The advertisement could be a billboard, a magazine or newspaper ad, or a TV commercial. Students should decide whether they are targeting the ad toward kids or teachers and parents.

Further Reading

Faughn, J.; Turk, J.; Turk, A. *Physical Science;* Saunders College: Philadelphia, 1991. (Teachers)

Kirkpatrick, L.; Wheeler, G. *Physics: A World View;* Saunders College: Philadelphia, 1992. (Teachers)

Zubrowski, B. *Raceways: Having Fun with Balls and Tracks;* William Morrow: New York, 1985. (Students)

Contributor

Anita Kroger, Gifted and Talented Specialist, Cincinnati, OH; Teaching Science with TOYS, 1986–87.

Handout Masters

Masters for the following handouts are provided:
- Observation Sheet
- Instructions for Small Group or Learning Center Investigation

Copy as needed for classroom use.

PHYSICS WITH A DARDA® COASTER

Observation Sheet

1. After having watched the demonstration, I think the most important factor in determining whether the car makes it around the loop is…

   _____

   _____

   _____

2. As the car goes up the loop, it (speeds up, slows down).

   _____

3. How could you get the car to go around the loop without winding the motor?

   _____

   _____

   _____

   _____

4. If you use this method, where does the car's initial energy come from?

   _____

   _____

   _____

   _____

5. How is this energy stored until it is needed?

   _____

   _____

   _____

   _____

   _____

PHYSICS WITH A DARDA® COASTER

Instructions for Small Group or Learning Center Investigation

If you are doing the Darda® coaster activity as a small group or learning center investigation, cut out the Loop-the-loop instructions (Steps 1–11), and mount them on index cards, one card per step. This is a good technique for presenting the instructions when you want to include information on a step that you don't want students to know at the outset. You may also want to prepare a worksheet on which the students can record their answers to the questions posed on the instruction cards.

If you are doing Extensions 1 and/or 2, cut out the Jumps (Steps 1–5) and/or Measurement (Steps 1–4) instructions.

NOTE ABOUT STEP 7: When/if the students come to you for permission, feel comfortable telling them to go on. The next card is not an answer but a hint.

Loop-the-loop

❶ Run your car on the floor or a table several times.

Loop-the-loop

❹ Why does the car sometimes go around the loop and sometimes not? What determines this? Discuss this question. Write down your answers and then share them. If you don't agree, discuss and try to come to an agreement.

Loop-the-loop

❷ Play with your car and track until you can make the car go around the loop several times.

Loop-the-loop

❺ Run the car so that it will go around the loop again. This time watch to see where its speed is increasing and where it is decreasing. You may need to do this several times. Record your observations.

Loop-the-loop

❸ Run your car into the loop in such a way that it will not go around the loop.

Loop-the-loop

❻ You should have determined that the car slowed down as it went up the loop. If this was not what you saw, stop and discuss your observations with your teacher before continuing. (If your teacher is busy, record your observations, assume the car slows down, and see your teacher as soon as possible.) Discuss why the car slows down as it goes up. Write down the reason on the sheet.

[Continued.]

Loop-the-loop

❼ You've been running the car around the loop by pulling it backward to wind its motor. Now find another way to make the car go around the loop without winding its motor or pushing it. This may be hard, but it can be done. After you come up with your idea, try it. DO NOT GO ON TO THE NEXT CARD UNTIL YOU COME UP WITH YOUR IDEA AND TRY IT, OR GET PERMISSION TO GO ON.

Loop-the-loop

SKIP IF YOU SOLVED THE MYSTERY OF CARD ❼.

❼a Think of a very tall roller coaster. Does this give you any ideas? Work on this awhile. If this doesn't help, do ❼b.

Loop-the-loop

❼b Lift up the beginning of the track.

Loop-the-loop

❽ Measure how high you must place the beginning of the track for this to work. Think and talk about why this works. Where is the car's energy coming from now? How is this different from what happens when you pull the car back?

Loop-the-loop

❾ Run the car by pulling it back, and have it run the loop unsuccessfully. Observe it carefully as it runs the track and falls and record your observations.

Loop-the-loop

❿ Run the car by lifting the track and have it run the loop unsuccessfully.

Loop-the-loop

⓫ Why did the car fall? Discuss this question. Did you notice that it does not stop before falling? What does this tell you? Write your answer on the sheet.

Jumps

❶ Run the car so that it successfully makes a jump. What makes the car able to do that? Discuss and record answers.

[Continued.]

Jumps

❷ What happens if the jump is made wider? Try jumps of various widths and record what happens.

Measurement

❶ Lay a meterstick on the floor. Put the car down beside the meterstick and, pressing down on the back of the car, pull it back 20 centimeters. Let it go. Mark where it stops and measure the distance. Record your observations.

Jumps

❸ What happens if one side is higher than the other? Try this with various heights and record what happens.

Measurement

❷ Repeat the above, pulling the car back 30, 40, 50, 60 centimeters. Keep going up if you'd like. Record your observations and note any patterns. Graph the results. Can you use this graph to make predictions? If so, write some predictions for distances you haven't tried and then try them out. Were you correct? Do several trials and average them. Is this answer closer to your prediction?

Jumps

❹ What happens if one side of the track is slanted? What if they're both slanted? What if they're both slanted similarly, or differently? Try and record your observations.

Measurement

❸ Do the same with running the car in the loop. Pull the car back 20 centimeters and see if it goes through the loop. If it does so, see how far it goes after it leaves the loop. Record your observations. Continue with distances of 30, 40, 50 centimeters, etc. As before, graph, predict, try it out.

Jumps

❺ Discuss the reasons for the observations you made and write down your comments.

Measurement

❹ Identify some length of pull-back distance that will get your car through the loop. Compare the results when you use a pull-back distance that length with the loop and just on the floor. Why do you think this happened? Write your ideas about this.

EXPLORING ENERGY
WITH AN EXPLORER GUN®

Students observe how an Explorer Gun® stores energy and works.

An Explorer Gun® with disk "ammunition"

GRADE LEVELS

Science activity appropriate for grades 5–8
Cross-Curricular Integration intended for grades 5–6

KEY SCIENCE TOPICS

- energy
- kinetic and potential energy
- work
- averages and ranges

STUDENT BACKGROUND

Students should be familiar with the concepts of work, potential energy, and kinetic energy. The Teaching Science with TOYS activity "Push-n-Go®" provides a good introduction to these concepts. Investigating the Explorer Gun® will provide the opportunity to extend students' understanding of these concepts.

Although not essential, some previous experience with projectile motion would be helpful in understanding this activity, particularly if the Variation or the Extension is used. This experience could be provided by doing the Teaching Science with TOYS activity "Two-Dimensional Motion."

KEY PROCESS SKILLS

- collecting data   Students collect data from repeated trials.

- making graphs   Students use their data to make bar or line graphs. Classroom data may be compiled and graphed.

TIME REQUIRED

Setup	10	minutes
Performance	60	minutes
Cleanup	5	minutes

Materials

For the "Procedure"
Per group
- Explorer Gun® by Park Plastic

See "Getting Ready" for details.

- meterstick, metric measuring tape, or trundle wheel

For "Class Discussion"
Per class
- a toy that utilizes a twisted (not compressed) spring, such as a See-Thru-Loco or the Press-N-Roll (by Lil Hands)

For the "Extension"
Per group
- paper
- scissors
- tape
- clay
- sheets of Styrofoam™

Safety and Disposal

Remind students not to fire the Explorer Gun® at anyone. No special disposal procedures are required.

Getting Ready

1. The Explorer Gun® is a toy that winds up and shoots a disk consisting of an outer rim and three helicopter-type blades. Several manufacturers sell similar items; however, the gun manufactured by Park Plastic is more durable. Even with the Park guns, it is useful to have an extra gun available in case one breaks during the experiment.

   Trial test each Explorer Gun® to determine the maximum number of half-turns the gun will allow. This varies from one gun to another but will be about seven. Be sure to warn the students not to force the gun beyond this number of half-turns. Measure the distance the disk travels with the maximum number of winds, and make sure the area to be used for the experiment is large enough to accommodate this distance.

2. The "Procedure" of this activity does not present step-by-step instructions for the experiment because the exact design is left up to you. As long as the number of half-turns of the disk is the variable and all other factors are held constant, a variety of experimental designs will yield good results. Ideally, students should test six different "numbers of half-turns."

   Since the Explorer Gun® is a toy rather than a precision scientific instrument, repeated measurements with the same gun will not give identical results. Thus, it is ideal for each group to do several trials for all six "numbers of half-turns" and average their results. However, this may be too time-consuming.

Alternatively, you could have each group test each "number of half-turns" once and then collect class data for averaging. (See the "Explanation" for further discussion.) Whether the groups do each "number of half-turns" once or several times, they should all do the same six "number of half-turns."

Procedure

Part A: Discussing Variables

1. Show the students the Explorer Gun® and demonstrate the proper method for firing it. Fire it several times, winding it a different "number of half-turns" each time. Explain to the students that a scientist might wish to investigate the relationship between the amount of winding and the distance flown. Announce that the purpose of the activity is to discover whether the disk will fly twice as far if it is wound twice as many times. Write the question, "Does winding the gun six half-turns cause the disk to travel twice as far as three half-turns?" on the board, or have a student do so. Explain the following ideas: "An experiment must be set up so that it can be exactly repeated numerous times. Part of making this experiment repeatable is keeping the gun consistently level and always firing from the same height." You may want your students to use a height of 1 m, since the top of a vertical meterstick can be used as a reference.

2. Demonstrate the method that you want students to use for counting half-turns used to wind the gun. One method is to count half-turns by placing the thumb on top of the disk and then turning until the thumb is on the bottom. Alternatively, you could count the number of "clicks" heard while winding. Or you could put a bright dot of paint on the disk and count the number of times the dot travels a complete circle.

Part B: Discussing the Experiment

1. Describe the experimental design you would like your students to use. (See "Getting Ready.")

2. Have the students name any conditions of the experiment that are being kept constant and any that are variables. List these on the board.

Part C: Conducting the Experiment

1. Break students into cooperative groups.

Each group will need at least one Firer, Recorder, Marker, and Measurer to perform the following jobs:
- *Firer—to turn the disk and fire the gun,*
- *Recorder—to watch to be sure the gun is level and record the data,*
- *Marker—to spot and mark the landing point, and*
- *Measurer—to measure the distance to the landing point.*

2. Distribute materials and move to the gym or outdoors.

Students may use metersticks, metric measuring tapes, or trundle wheels to measure distance.

3. Have students carry out experiments and collect data.

Part D: Compiling Data

1. Upon return to the classroom, each group should use its data to make line or bar graphs of the distance traveled versus the "number of half-turns." If each did several trials for each "number of half-turns," each group should average its own results before making the graph. If students are making line graphs, remind them not to connect the dots. (See Figure 1.)

Figure 1: Effect of Winding on Projectile Range of Explorer Gun

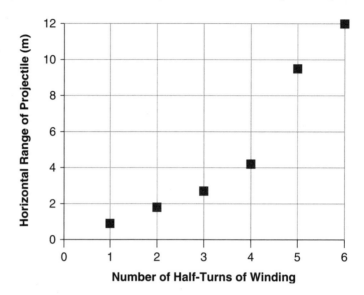

2. One student from each group should put the group's data on the board. If one group's data stand out as quite different from the rest, check their list of variables and constants and have them demonstrate their procedure. Students should then average all the values for a given "number of half-turns."

3. (optional) A graph using these average values should be drawn on the board. Although this step is optional, it provides an excellent platform for introducing or reviewing measurement uncertainties. (See the "Explanation.")

Class Discussion

Lead the students through a discussion of how energy is stored by the gun and then converted into kinetic energy when the trigger is pulled. Since students may not be familiar with springs that are twisted rather than compressed, it is useful to have such a spring to show the students. One source is a See-Thru-Loco that operates on the same principle and has a clear plastic body, so the spring and gears can be seen. Another is a Crazy Wheel. The tire is easily removed from the wheel so the spring is visible.

You should also make the connection between the initial velocity of the disk and the distance it travels before hitting the ground. This should be a review of ideas from previous studies of motion.

Return to the question that initiated the experiment, "Does winding the gun six half-turns cause the disk to travel twice as far as three half-turns?" (You can use either the raw data or the graph.) The points on the graph will not fall in a straight line, and the answer is clearly no.

To illustrate that the first few half-turns do not store much energy, compare the increase in distance traveled in going from one to two half-turns with the increase from six to seven half-turns. The students should recall that when winding the gun they had to exert a larger force near the end than at the beginning. Make sure the students understand the connection between the size of the force and the amount of energy stored.

If you did a graph with the class averages, you will also want to discuss why this is better than using just one trial. Use the graph to predict the distance traveled for some number of half-turns that you did not measure. Have one group make this measurement at a later time and report back to the class. This illustrates the usefulness of the graph and ties back into the discussion of the meaning of average as a "best guess."

Explanation

 The following explanation is intended for the teacher's information. Modify the explanation for students as required.

As you wind the Explorer Gun®, you do work. Energy from your body is stored in the spring as elastic potential energy and then later converted into the kinetic energy of the disk.

You do work as you wind the gun because you apply a force to the disk and the point where the force is applied moves as a result. The work required for the first turn of the disk is not the same as that required for the fourth because more force is required with each turn. When you turn the disk, a spring is twisted, as in watches that use mechanical energy rather than batteries. The spring stores the energy. The amount of work done, and thus the amount of energy stored, depends on the total number of half-turns. The relationship between the number of half-turns and the amount of energy stored is not linear because the force required for each turn is not constant.

When you pull the trigger of the gun, the spring untwists, turning the axle to which the disk is attached and pushing the disk forward off the axle. The amount of stored energy in the spring (that is turned into kinetic energy) determines the velocity of the disk when it leaves the gun. Although it is not necessary to raise the issue with the students, you should understand that this is a more complicated case than, for example, a dart gun that stores energy by compressing a spring. Part of the energy stored in the Explorer Gun® goes into kinetic energy associated with the rotation of the disk rather than just into the kinetic energy associated with the forward motion as in the dart gun.

The velocity of the disk as it leaves the gun determines the distance it travels before hitting the ground. As soon as the disk leaves the gun, it begins to fall due to the force of gravity. The amount of time it takes to drop is a function only of the height from which it begins, which will be the same for all trials. The distance the disk travels before hitting the ground is the product of this amount of time in the air and the initial velocity. A more complete discussion of these ideas may be found in the Teaching Science with TOYS activity "Two-Dimensional Motion."

Although the "number of half-turns" is supposed to be the only variable in the experiment, the springs in the various guns are not identical and may also be a variable, depending on how you compile the data. For example, if each group did several trials for each "number of half-turns," and averaged and graphed ONLY their own data, the spring is a constant for that group of data because all the data were generated using the same gun. But if each group collected only one value for each "number of half-turns" and all the data were used to create a class average, the spring becomes a variable because the data used to create the class average were generated using different guns with different springs. If all the groups then combine their averages into a class average, the spring is again a variable for the same reason as before.

Good experiments generate data that allow experimenters to make predictions. The average values generated from many trials are a "best guess" prediction of the data that would be generated if the experiment were repeated. The certainty of this "best guess" depends, in part, on the number of values collected, the range (span between smallest and largest), and how closely these values are clustered near the average. If we have very few values, and they have a very wide range, we might question the usefulness of the average of those data as a predictor. If we have many values, and they still have a wide range, but many are clustered near the average value, we may feel more confident in the average of those data as a predictor. If we have many values, they have a narrower range, and many are clustered near the average, we may feel even more confident in this average as a predictor. (See Figure 2.)

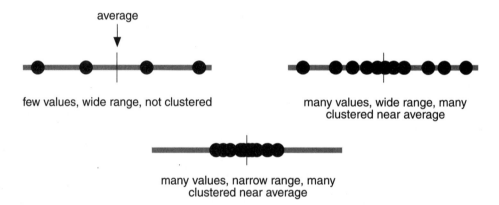

few values, wide range, not clustered

many values, wide range, many clustered near average

many values, narrow range, many clustered near average

Figure 2: The average of a set of values is a more reliable predictor when many values are clusterd in a narrow range near the average.

Hidden variables (such as the spring or wind) can make data more scattered and less reliable. Usually, the more data you collect, the less impact a hidden variable like the spring will have on the reliability of your averages. If each group did several trials for each "number of half-turns" and therefore you have many more values when computing your averages, these averages should be more reliable predictors of the values you would get if you repeated the experiment. The range in values for a given "number of half-turns" and how many of these values are fairly close to the average will give you an idea about the reliability of your averages as predictors. The smaller the range, the closer you would expect the new measurement to be to the average. Your students can test the reliability of one of your averages by repeating part of the experiment using a randomly chosen gun and then comparing the new result to the average.

Variation

Instead of the activity described above, a more general investigation could be done. Ask the class, "How can we get the disk to go the greatest possible distance?" After brainstorming a little, divide into three groups: one to do the winding experiment, one to investigate the effect of the firing angle, and the third to investigate the effect of height above the ground. Caution the students to keep everything constant except the one variable they are investigating.

The data can be presented in tables or graphs. The follow-up discussion should include how to combine all three results to maximize the range. For further discussion of these results see the Teaching Science with TOYS activity "Two-Dimensional Motion."

Extension

Pose the question, "How could we change the design of the gun to make the disk go farther (or not as far)?" Brainstorming will probably produce several ideas that can be checked out and several that cannot. Possibilities that cannot easily be tried are to use a spring that winds more easily, take out the blades, and make it easier to pull the trigger.

Some ideas that can be checked by experimentation are making the disk heavier, using a solid disk, and increasing the diameter of the disk. Groups should first make a measurement using the original disk to compare to the results after the changes are made. Various amounts of clay can then be added to the rim of the disk, being careful to distribute it evenly so the disk is still balanced. Paper can be taped over the front of the disk to eliminate the effect of the blades and simulate a solid disk. If you have access to sheets of a lightweight material like Styrofoam™, you could try increasing the diameter of the disk by taping a Styrofoam™ ring outside the disk.

There are a number of important aspects to this extension. Foremost, the students are actively involved in "What will happen if…" thinking. They must decide how to test their ideas and carry out the experiment. Discussion of the results is also important. Encourage the students to speculate about why the results were as they were. Do not feel that you must be able to give a scientific explanation in each case. This discussion might be used as a springboard into further reading on a variety of topics including aerodynamics and spinning objects.

Assessment

The following is a list of possible assessment questions. You might want to use some of them as group discussion questions, journal writing assignments, or on a written examination.

1. When firing the Explorer Gun®, why does the disk go so much farther with six half-turns than with two half-turns?

2. Would you expect the disks to go farther if the gun was straight when fired or if the gun was tilted up slightly? Why?

3. When does the Explorer Gun® have the greatest elastic potential energy?

4. Give an example of a change of energy from one form to another in the Explorer Gun®.

5. Why is it good procedure to take several measurements and average them?

Cross-Curricular Integration

Art:
- Have students draw, paint, or make clay objects that store energy. A group collage of energy-storing objects would be an interesting class project.
- Students may also investigate what is meant by the phrase "kinetic art," and find pictures of it or even construct some.

Language arts:
- Once the students are introduced to the topic of potential versus kinetic energy, they may begin to notice the wide variety of objects in the world that store energy and the variety of ways in which that energy is released. Use students' observations about the work done by a plant growing up through a crack in the pavement, extending the crack further, and crumbling the concrete to conduct a discussion.
- Conduct a brainstorming activity in which students mention as many energy-storing objects as possible as a springboard for an writing activity. Use the list to inspire creative writing.
- Ask students to search poetry books for poems that describe objects that store and release energy.

Math:
- Averaging—If your students have not yet been introduced to the technique of averaging, you may even wish to conduct this lesson as a math activity rather than a science activity, particularly if your science time is limited, in order to give the students a concrete example of averaging.

Social studies:
- Tie the discussion of uncertainties in measurements to opinion polls and their uncertainties.

Further Reading

Frances, N. *Super Flyers;* Addison-Wesley: Reading, MA, 1988. (Students)

Gartrell, J. E., Jr.; Schafer, L. E. *Evidence of Energy;* National Science Teachers Association: Washington, D. C., 1990. (Teachers)

Kirkpatrick, L.; Wheeler, G. *Physics: A World View;* Saunders College: Philadelphia, 1992; pp 117–132. (Teachers)

Walpole, B. *Fun with Science: Movement;* Warwick: New York, 1987. (Students)

BOUNCEABILITY

*Students investigate the bouncing of balls
to determine why some are better bouncers than others.*

A racquetball on carpet

GRADE LEVELS

Science activity appropriate for grades 4–7
Cross-Curricular Integration intended for grades 4–6

KEY SCIENCE TOPICS

- elastic potential energy
- gravitational potential energy
- kinetic and potential (stored) energy

STUDENT BACKGROUND

This activity should not be used as an introductory lesson on
energy. Students should be familiar with kinetic energy, elastic
potential energy (springs and rubber bands), gravitational
potential energy, and heat. They should have investigated other
stored energy toys in Teaching Science with TOYS activities such
as "Push-n-Go®" and "Physics with a Darda® Coaster."

KEY PROCESS SKILLS

• measuring	Students measure the heights of ball bounces.
• making graphs	Students graph their data.
• controlling variables	Students plan an experiment to assess the bounceability of various balls.

TIME REQUIRED

Setup	5	minutes
Performance	80	minutes
Cleanup	5	minutes

Materials

For the "Procedure"
Per group of 3–5 students
- 1 each of a variety of balls such as tennis, golf, rubber, Styrofoam™, hi-bounce, metal, clay, and racquetball

- a variety of surfaces to bounce the balls on, such as carpet, grass, flooring tiles, ceiling tiles, metal plate, cardboard, cork, foam pad, and Styrofoam™

The exact number of balls or surfaces is not important.

- 1 meterstick
- chart paper for graphs and colored markers
- (optional) construction paper

For "Variations and Extensions"

❶ All materials listed for the "Procedure" plus the following:
- access to a refrigerator or radiator (or other heater)

❷ Per class
- 1 Hoppity Popper or racquetball with top third cut off

Hoppity Popper is available from American Science and Surplus, P.O. Box 48838, Niles, IL 60714-0838.

❸ Per class
- 1 Smart Ball™ (by Applied Elastomerics), Splat Ball, or Jelly Ball

These are all basically the same toy by different manufacturers.

Safety and Disposal

No special safety or disposal procedures are required.

Getting Ready

1. (optional) Ask students to bring balls from home.

2. Put numbered lists of the balls and surfaces that will be used on the board.

3. For young students, determining how high the ball bounced in centimeters may be difficult. If this is the case, mark the metersticks off into 10-cm sections and wrap each section in a different color of construction paper. The students can then record the ball's bounce height as "blue," "yellow," or some other color, depending upon which color section it bounces into.

Introducing the Activity

Remind the students that they have already investigated several "stored-energy" toys. Review the concepts of kinetic and potential energy using one or more of these toys. For instance, demonstrate a spring-up toy and have the students describe the energy transformations that take place.

Present the following ideas to your students: "A ball would seem to be the simplest possible stored-energy toy; however, we did not begin with it because it is really a lot more complicated than you might think. Today we want to investigate the bouncing of balls to try to determine why some are better bouncers than others."

Part A: Discussing the Variables

1. Show the students the balls they will be testing. Ask them to think about which one will be the best bouncer, then take a vote. Randomly select a few students who voted for different balls. Ask each of these students to explain to the class his or her reasons for selecting a particular ball.

2. Ask the students what variables might affect how high a ball bounces. Some ideas may have come out as students explained their choices in Step 1. If so, write these on the board, then encourage the students to brainstorm other ideas. Possible variables include the material the ball is made of, its temperature, whether it is dropped or thrown, the height from which it is dropped, and the surface that it hits. At this point, do not try to limit the students to ideas that you think are "reasonable."

Part B: Planning the Experiment

 If your class does not have prior experience with planning experiments, you could develop the plan through a whole group discussion, rather than through small group work.

1. Tell the students that today you are going to investigate two of the variables that affect how high a ball bounces: the material the ball is made of and the surface on which it is bounced. Tell the students that the first step is to design an experiment that allows them to test both of these variables.

2. Divide the students into groups of three to five. Ask each group to figure out how they could do an experiment to determine which surface makes the ball bounce highest. Depending on their prior experience, you may want to remind them to control variables.

3. Once each group has completed their plan, select one group to explain their plan to the class. Get the rest of the class to critique it. Is it a good plan? Have they forgotten to control one of the variables?

 A reasonable experimental plan is to drop each ball from a standard height (such as 1 m) measured from a standard position on the ball (say, the bottom), and then "eyeball" the rebound height when the ball bounces. Whatever dropping height and measuring position on the ball you choose, make sure that all students are consistent.

4. After one good plan is established, ask if any other group has a different idea.

5. If several different plans are proposed, explain to the class that they must choose one plan for everyone to follow so that they can then compare data afterwards.

Part C: Conducting the Experiment

Students should conduct the experiment in their small groups. Each group will need a Ball Dropper, an Initial Height Checker, and one or several Rebound Height Checkers to perform the following jobs:

- *Ball Dropper—to hold the meterstick and drop the ball;*
- *Initial Height Checker—to stand back and make sure the initial height is right; and*
- *Rebound Height Checkers—to note the rebound height.*

A possible way to collect a large quantity of data is as follows:

1. Set up the various surfaces to be tested at different locations around the room.

2. If sufficient wall space is available, tape metersticks to the wall, or tape lengths of paper to the wall so that the Rebound Checkers can mark the paper and measure the height.

3. Assign each group a different kind of ball (or two balls if you have a sufficient quantity of different ones).

4. Have the groups rotate through the locations with different surfaces, testing their ball(s) at each stop.

5. Instruct each group to make a bar graph of their data on chart paper large enough for the whole class to see. They should have a bar for each surface, and the height of the bars should show the height achieved by the bouncing ball. To facilitate comparisons, each group should use a vertical scale of 0 to 100 cm. Have groups post graphs on the wall. For sample graphs, see the Sample Graphs sheet (provided). This sheet is for your reference only; students should make their own graphs.

Make sure all groups put the surfaces on the chart in the same order to facilitate comparisons. Color-coding can be used to increase readability. For instance, all bars representing carpet might be green and all bars representing Styrofoam™ yellow. Colored markers can be used or strips of construction paper cut in advance which can then be trimmed to the appropriate length and taped on the chart.

Class Discussion

Whether you do most of the talking or the students do will depend on their background in energy concepts.

Explain to students that they have clearly demonstrated that not all balls bounce equally and that the surface does indeed have an effect on how high the ball bounces. Discuss the energy transformations that take place as the ball falls and then bounces back. Ask students what kind of energy the ball has before you drop it, while it is falling, and just before it hits the floor. "Why doesn't the ball come back to its original position? Did it lose energy?" (See the "Explanation.")

Now help students apply these ideas to understanding the data on the graphs. Which ball is the best bouncer overall? Remind students that only balls dropped on the same surface should be compared. Saying that a tennis ball dropped on carpet bounces better than a Ping-Pong™ ball dropped on rubber padding doesn't tell you much.

Teaching Physics with TOYS

Some of these observations may emerge from your discussion:

- Students may note that the hi-bounce ball is rarely the best bouncer on any surface, but is reasonably consistent from surface to surface.

- Ask students to look at how each of the balls bounced on the foam and the metal. They should observe little variation from one ball to another on the foam. You can explain that the foam deforms easily, so the stiffness of the ball doesn't matter much. Students should observe the biggest variation on the metal plate, because it deforms very little.

- Students may observe that the clay bounces significantly only on the foam and does not bounce at all on the metal and floor tile. The clay loses its kinetic energy in a permanent shape change when it hits hard surfaces, so it can't bounce back up.

Continue the discussion as long as the students are interested in thinking about the reasons why different kinds of balls bounced differently on the test surfaces.

Explanation

 The following explanation is intended for the teacher's information. Modify the explanation for students as required.

Before the ball is dropped, it has an amount of gravitational potential energy that depends on its height above the floor. As the ball falls, this potential energy is gradually turned into kinetic energy. The greater the ball's speed as it falls, the greater the kinetic energy. On the way back up, the kinetic energy is being turned back into potential energy. At the top of the bounce, the ball's energy is once again all potential energy. Since the ball does not bounce back to its starting height, it has less potential energy now.

Since energy cannot be "lost," the missing energy must go into some other form of energy. The energy transfer happens during the interaction between the ball and the surface. Part of the energy goes into the sound waves that are produced when the ball hits the surface. Part of the energy is converted to heat in the ball and in the surface. You can observe this if you drop a ball of clay several times in rapid succession. The clay will begin to feel warmer.

What determines how high different balls bounce on the same surface? Much of the difference is a result of how much the balls deform and even more importantly how fast they "un-deform." During the bounce, the shape of the ball changes. This shape change takes energy, just as stretching a rubber band does. Flattening the ball is similar to compressing a spring. You get the energy of compression back as the shape goes back to normal. The clay doesn't bounce well because it stays deformed. If the ball is still partially deformed after it leaves the floor (or other surface), the energy that was stored in that deformation does not return to kinetic energy of the ball even though the ball does later return to its original shape.

What determines how high the balls bounce on different surfaces? During the bounce, both the shape of the ball and the shape of the surface are deformed. The height of the bounce is determined by how much energy of compression is returned as the shape of both the ball and the surface go back to normal. Each ball type and surface type interact differently, producing a unique result. Even so, some surfaces produce fairly consistent results with all types of balls. For example, all the balls bounce in the foam, because the foam deforms rather than the ball, acting much like a trampoline. In contrast, if the surface stays deformed as the Styrofoam™ may, then the energy that went into causing the deformation does not return to the ball.

Extensions

1. Test the prediction that temperature affects bouncing by cooling several balls in the refrigerator overnight or heating them by a radiator or other heater. Measure how high the heated or cooled balls bounce and compare this data with the "room-temperature" measurement.

2. Ask the class if there is any way a dropped (not thrown) ball can bounce back up higher than where it started. Demonstrate the Hoppity Popper or cut-off racquetball (see photo) several times. (Turn it inside out and drop it concave side down.) Ask for possible explanations of what is happening. After discussing students' ideas, explain the following: "When you turn the Hoppity Popper inside out, you are storing energy in it just like stretching a spring. When the Hoppity Popper hits the floor, it returns to its original shape and the energy you stored is converted to kinetic energy. Thus, as the kinetic energy is changed back into gravitational potential energy, it has enough energy to surpass its initial height."

3. Balls made of a soft polymer are marketed as Splat Balls or Smart Balls™. If thrown at the floor, these spread out almost completely flat then slowly reform. This is a very visible example of the deformation of the ball when it hits the floor.

    If your ball stops working properly, wash it with soap.

Assessment

This assessment can be done either individually (written) or as a group (oral). If done as a group assessment, explain that all members of the group are responsible for the "answer," as any one of them could be asked by the teacher to respond.

Ponder one of the following questions:

1. Explain what playing basketball on a carpeted court would be like.

2. You and your friends are all set to play baseball, but you can't find the ball. Your little brother comes up with a tennis ball. You decide to try it. How will your game be different?

3. Some tennis tournaments are played on grass. How do you think those games are different from those played on a paved court?

Cross-Curricular Integration

Language arts:
- Read and discuss the book *The Pinballs,* by Betsy Byars (Scott Foresman, ISBN 0673801365). In the book, one of the foster children describes herself and the others as pinballs—they bounce from here to there, wholly dependent on those around them.
- Have students imagine what it would be like to be a certain type of ball and explain in writing how they would react to different surfaces and how the surfaces would affect their bouncing.
- Have the students invent and describe a brand new game using one of the tested balls and a surface other than the one it is typically used on.

Math:
- The activity involves measurement, graphing, and charting.

Physical education:
- Discuss why different kinds of balls are used in different sports.

Further Reading

Doherty, P. "That's the Way the Ball Bounces," *Exploratorium Quarterly.* Fall 1991. (Teachers)

Turk, J.; Faughn, J.; Turk, A. *Physical Science;* Saunders College: Philadelphia, 1991; pp 59–66. (Teachers)

Kaplan, N. "But That's Not Fair," *Science and Children;* April 1984. (Teachers)

Kirkpatrick, L.; Wheeler, G. *Physics: A World View;* Saunders College: Philadelphia, 1992; pp 117–132. (Teachers)

Radford, D. *Science from Toys;* Macdonald Educational: London, 1972; Chapter 7, Science 5/13. (Teachers)

Stone, H.; Siegel, B. *Have a Ball;* Prentice-Hall: Englewood Cliffs, NJ, 1969. (Students)

Contributors

Anita Kroger, Gifted and Talented Specialist, Cincinnati, OH; Teaching Science with TOYS, 1986–87.
Bruce Peters, Center for Chemical Education, Middletown, OH.

BOUNCEABILITY

Sample Graphs

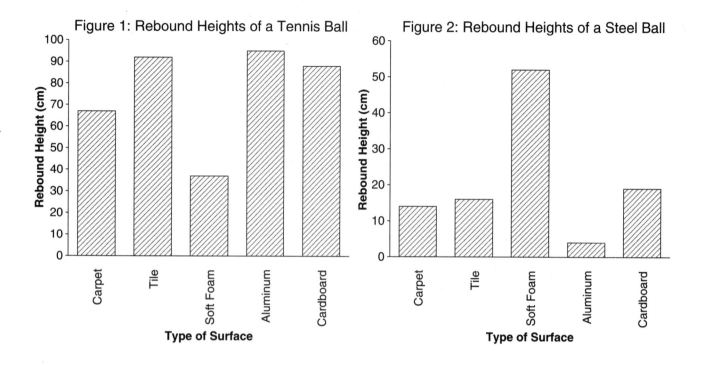

Figure 1: Rebound Heights of a Tennis Ball

Figure 2: Rebound Heights of a Steel Ball

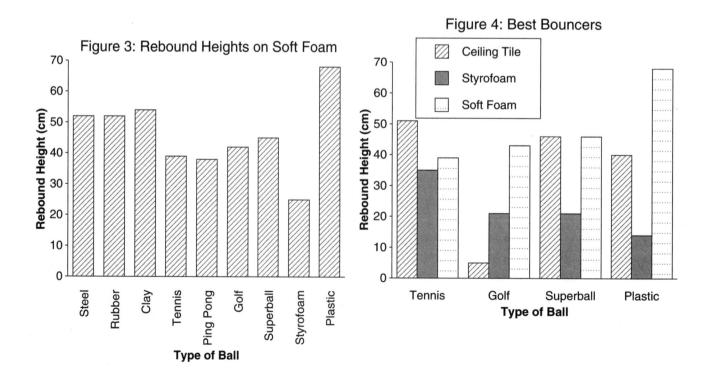

Figure 3: Rebound Heights on Soft Foam

Figure 4: Best Bouncers

ENERGY TOYS LEARNING CENTER

Students observe a variety of ways in which energy can be stored and released and apply knowledge about kinetic and potential energy to stored-energy toys.

Energy-storing toys

GRADE LEVELS

Science activity appropriate for grades 5–6
Cross-Curricular Integration intended for grades 5–6

KEY SCIENCE TOPICS

- elastic potential energy
- gravitational potential energy
- kinetic energy
- work
- forces

STUDENT BACKGROUND

Before beginning the explorations in the center, students should be very familiar with the concepts of work, forces, energy, kinetic energy, and potential or stored energy. The learning center will give the students an opportunity to review, apply, and further explore these concepts. Students should have seen or done a couple of explorations similar to those in the center. If a toy has already been seen by the entire class, do not include it in the learning center.

KEY PROCESS SKILLS

• communicating	Students record their observations and give written explanations of kinetic and potential energy based on their observations.
• investigating	Students investigate kinetic and potential energy of toys in a learning center.

TIME REQUIRED

Setup	45	minutes
Performance	60	minutes (over several days)
Cleanup	45	minutes

Materials

Per class:
- toys that store and release energy—some possibilities are

Spring-ups	Flip Frog
Push-n-Go®	Popovers®
See-Thru-Loco	Bath Tubbie
Turblo	Jumping Disks
Explorer Gun®	dart gun
High Flyer	pull-back car
Ping-Pong™ gun	any battery-operated toy

Safety and Disposal

No special safety or disposal procedures are required.

Introducing the Activity

Explain to the students that, over a period of days (or perhaps a week), they will choose three toys from the center, play with and observe them, and then record their observations and answer the questions on the Data Sheet (provided). Point out the toys and worksheets in the center. You may wish to demonstrate any unusual toys.

Procedure

As each student takes a turn in the learning center, have her or him do the following:

1. Choose a toy from the center, get a Data Sheet, and investigate the toy.

2. Record observations and answer the questions on the Data Sheet.

3. Turn in the Data Sheet.

4. Repeat Steps 1–3 with at least two other toys.

> *Students may choose more toys if they have time, but should not use more than three toys until the sheets from the first three have been turned in.*

Class Discussion

After the center explorations are complete, hold a class discussion to help students summarize their observations. The focus of the discussion will be the patterns in observations of the energy toys and the generalizations which may be made about these observations. Briefly demonstrate and discuss each toy so that all students may see all toys.

> *Since this discussion is presented as a review of concepts, one class period or less should suffice.*

The following explanation is intended for the teacher's information. Modify the explanation for students as required. You may want to read the Data Sheet before reading this explanation.

Obviously, a specific explanation cannot be provided for every toy that could possibly be included in a learning center. However, some explanations that can easily be adapted to a variety of toys are provided below.

Explorer Gun

The kinetic energy you observe with this toy is in the motion of the flying disk. This energy is supplied by your muscles turning the disk when you attach it to the gun. This energy is stored as elastic potential energy in a spring in the toy. It stays there until you pull the trigger, which allows the spring to unwind. The energy stored in the spring is converted to the kinetic energy you observe in the disk's flight. You can change the amount of kinetic energy from one trial to another by winding the disk more or fewer times. If you give the toy more energy, more energy is transferred to the disk. The toy has the most elastic potential energy when you have wound the disk as far as you are going to wind it. When the disk has the most gravitational potential energy depends on whether you fire it horizontally or at an angle. If fired horizontally, the disk has the most gravitational potential energy at the very beginning of the flight. If fired at an angle, it has the most gravitational potential energy when it is at its highest point.

Spring-Ups

Before the toy begins moving, energy is stored as elastic potential energy in the toy's spring when the energy of your muscles pushes down on the toy and makes the suction cup stick. When the suction cup lets loose, the elastic potential energy is converted to kinetic energy. The toy has the most kinetic energy when the spring is completely expanded. You cannot change the height of the jump or the toy's speed from one trial to the next; the amount of kinetic energy is constant. As it moves up, the toy slows because gravity is pulling it down. As this happens, the kinetic energy is being changed into gravitational potential energy.

Spring-ups often have a flipping motion because the spring often bends slightly as the suction cup releases. As a result, the force exerted is not perfectly vertical.

At its highest point, almost all of the toy's kinetic energy is converted into gravitational potential energy. In most cases, the toy will still have some rotational motion, so it must still have some kinetic energy. As the toy comes back down, the gravitational potential energy is converted back to kinetic energy. When the toy hits the table and stops, it loses both its potential and kinetic energy. Where does the energy go? Primarily, it becomes heat energy, but some of it goes into sound energy.

Flip Frog/Popover

The frog moves, so it must have kinetic energy. This energy comes from your muscles when you press down on the frog's head. This muscle energy is then stored in the spring in the frog as elastic potential energy. The energy is stored there until the suction cup comes loose. This causes the elastic potential energy in the spring to be converted to kinetic energy, the energy of the frog's jump. You cannot adjust the amount of kinetic energy from one trial to the next. The kinetic energy is determined by the spring and the suction cup. This toy has the most kinetic energy when it first begins to jump. It has the most elastic potential energy when you finish pushing it down and attaching it to the base (before it jumps).

The motion of the Popover is similar. Instead of stretching a spring, you put energy in by winding a spring. As this spring unwinds, another spring is stretched just as in the frog. Instead of a suction cup release to initiate the flip, a gear mechanism inside the Popover releases when the spring has been stretched a certain amount.

Both toys' motion is in two parts: up and down, and backwards in a vertical circle (the flip). The up-and-down motion can be explained just like the Spring-Up's motion. The rotational motion is a little harder to explain. When the spring is released, it moves inward at both ends, pulling the head backward and the legs upward. Since the legs are off-center, this causes the toes to push down on the table, and in an example of action-reaction forces, the table in turn exerts an upward force on the toes. This causes the gorilla to flip backwards, turning around its center of mass.

High Flyer
This toy is excellent for illustrating the scientific definition of work (force times distance). As you pull the string, you can clearly feel the reaction force pulling back. Your pulling force, exerted as the string moves, does work that is then transferred to energy of the Flyer. Most of this energy is given directly to the flying disk, but some is stored in a twisted rubber band in the handle. When you release the cord, this stored energy is used to pull the cord back into the handle. The greater the distance you pull the string before releasing, the greater the energy of the disk. The moving disk has both kinetic energy and gravitational potential energy. It has the most gravitational potential energy when it is at the highest point in the flight. It has the most kinetic energy when its speed is greatest, but it is difficult to pinpoint exactly when that is. It might be as it leaves the launcher or it might be just before it hits the ground.

Jumping Disks
The disk has kinetic energy any time it is moving. This energy comes from you when you use your muscles to press the disk down and from heat in your hands, which causes the metal to expand. The energy is stored in the disk for a while, because of the properties of the metal itself. (It has sort of a springiness to it.) When the disk pops up, the elastic potential energy is converted to kinetic energy. It slows as it moves upward because the kinetic energy is being changed to potential energy. The disk has the most gravitational potential energy when it is at the top of its jump. As gravity pulls the disk downward, the potential energy is changed back to kinetic energy. There are also action-reaction forces involved. When the disk returns to is original shape it actually presses down on the table. The table pushes back, which causes the disk to jump.

Push-n-Go®

When you play with this toy, it moves forward. This is an example of kinetic energy. Another example of kinetic energy is the upward movement of the head after you remove your finger. The energy comes originally from you, as you push down on the head. You exert a force while moving the head, so work is done transferring energy from you to the toy. The energy is stored as elastic or spring potential energy for a moment in the spring inside the toy. When you let go, the spring extends, and its potential energy is converted to kinetic energy. The toy has the most kinetic energy immediately after the head returns to its original position. The toy has the most elastic potential energy when you have pushed the head all the way down and haven't yet released it. Since the toy never leaves the ground, it has no gravitational potential energy. Eventually the toy does stop moving, so it no longer has any kinetic energy. The force of friction between the wheels and floor causes the toy to stop. As part of this process, the toy's kinetic energy is converted to thermal energy. (Further explanation related to this toy can be found in the Teaching Science with TOYS activity "Push-n-Go®.")

See-Thru-Loco

The energy is supplied by human muscles winding the spring. This energy is stored in the spring as elastic potential energy and is stored there until you release the key. Then the potential energy is converted to the kinetic energy of the toy's movement. The entire locomotive moves, and its internal parts also move. Both of these movements involve kinetic energy. This toy has the most potential energy when you have finished winding the key and haven't yet released it. The toy has the most kinetic energy when it is moving fastest—somewhere in the middle of the motion. The force of friction between the tires and the floor causes the toy to slow down and eventually stop. The toy's kinetic energy is turned into thermal energy—the toy and the floor each get a little warmer.

The gears transfer the energy from the spring to the wheels of the locomotive. The function of each gear should be identified. Notice that different gears turn at different speeds. To see the gears turn in slow motion, turn the winding key backwards. (Further explanation related to the operation of gears can be found in the Teaching Science with TOYS activity "Gear Up with a Lego® Heli-Tractor.")

Battery-operated toys

Although many different types of battery-operated toys are available, some general statements can be made about all of them. Batteries are storage devices for chemical energy. When the battery is placed in a complete circuit, the chemical energy is gradually converted to electrical energy. This electrical energy may be used to cause motion or produce light or sound or a variety of other forms of energy.

Variation

Divide the students into small groups and assign each group a toy to demonstrate and explain. Then have the class evaluate each group's presentation by answering the following questions:

1. Did the group explain all of the energy forms used? Explain.
2. Were you able to understand their explanations?
3. On a scale of 1 to 5, with 5 being the highest rank, how would you rank this presentation? Make comments if needed.

Cross-Curricular Integration

Earth science:
- After observing in detail how toys use and convert energy, students may be much more perceptive in recognizing energy use by humans, energy transfer, and energy waste. Present an ecology lesson regarding the ways and reasons why humans use energy either as a follow-up to this lesson, or as an introduction to the next area of study. To introduce the ecology lesson, have the students do a two-hour energy diary, recording ways in which they have used energy from within their own bodies and from other sources.

Language arts:
- The comparisons of energy toys lend themselves to writing exercises involving comparison and contrast.
- Have students write stories about what would happen if one of the energy toys that resembles an animal or person came to life or could talk.
- Students could write a persuasive piece—perhaps a radio or TV commercial—encouraging fellow students to use toys whose energy comes from the person playing with it, rather than from batteries.

Social studies:
- Have students study life in an earlier time, such as the pioneer era, ancient Rome, or the Middle Ages, and compare typical toys from that time to typical toys today. Encourage them to note how these historical toys stored and released energy.

Further Reading

Faughn, J.; Turk, J.; Turk, A. *Physical Science;* Saunders College: Philadelphia, 1991; pp 59–66. (Teachers)

Gartrell, J.E.; Schafer, L.E. *Evidence of Energy;* National Science Teachers Association: Washington, D.C., 1990. (Teachers)

Kirkpatrick, L.; Wheeler, G. *Physics: A World View;* Saunders College: Philadelphia, 1992; pp 117–132. (Teachers)

Contributor

Anita Kroger, Gifted and Talented Specialist, Cincinnati, OH; Teaching Science with TOYS, 1988–89.

Handout Master

A master for the following handout is provided:
- Data Sheet

Copy as needed for classroom use.

Name _____

Toy used _____ Date _____

ENERGY TOYS LEARNING CENTER
Data Sheet

Use the toy several times and observe carefully before answering these questions.

1. Describe in your own words what this toy does.

   _____

   _____

   _____

2. Does the toy change speed or direction? Yes _____ No _____
   If so, what caused it to do that?

   _____

   _____

   _____

3. What example of kinetic energy do you observe?

   _____

   _____

   _____

4. Where did this energy come from? (For example, batteries that store chemical energy.)

   _____

   _____

   _____

5. An object has the most kinetic energy when it is moving the fastest. At what point in its motion does this toy have the most kinetic energy?

   _____

   _____

   _____

[Continued.]

6. Can you make the amount of kinetic energy change from one trial to the next? If so, how?

_____

_____

_____

7. Was energy stored in this toy for a while? If so, where was it stored or how was it stored? Was this an example of gravitational potential energy or elastic potential energy?

_____

_____

_____

If your answer to #7 was No, skip questions 8, 9, and 10.

8. What caused the stored energy to be converted to kinetic energy?

_____

_____

_____

9. When does the toy have the most elastic potential energy? (Skip the question if it does not have any.)

_____

_____

_____

10. When does this toy have the most gravitational potential energy? (Skip this question if it does not have any.)

Use the space below to record any observations you made about this toy that are not included in your answers to the Data Sheet questions.

_____

_____

_____

_____

_____

SIMPLE MACHINES WITH LEGO®

Students use fun and popular Legos® to build working simple machines.

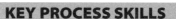

Lever built with Legos®

GRADE LEVELS

Science activity appropriate for grades 5–8
Cross-Curricular Integration intended for grades 5–6

KEY SCIENCE TOPICS

- (for some levels) definition of work
- simple machines

STUDENT BACKGROUND

Students should be familiar with levers, pulleys, gears, and wheels. Familiarity with screws and inclined planes is not necessary, because they can't be built effectively with Legos®. At least some students in each group should have prior experience building with Legos®.

KEY PROCESS SKILLS

- communicating   Students draw and describe the Lego® models they build as they perfect a model to incorporate as many simple machines as possible.

- classifying   Students classify Lego® models as to the type of simple machine represented.

TIME REQUIRED

Setup	5	minutes
Performance	40	minutes
Cleanup	5	minutes

Materials

For "Introducing the Activity"
- pictures of familiar objects that incorporate or are simple machines

For the "Procedure"
Per group
- 1 Lego® Technic Universal Set #8024 or #8022 or any larger Lego® Technic set
- string for pulleys
- (optional) pennies or paper clips

For the "Extension"
Per class
- uniform large masses, such as bricks
- seesaw

Safety and Disposal

No special safety or disposal procedures are required.

Introducing the Activity

Review levers, pulleys, and gears by showing pictures of familiar objects that are or incorporate simple machines and discussing how they work. Pictures might include a seesaw, a wheelbarrow, a truck loading ramp, and a bicycle.

Procedure

If sufficient Legos® are available, Steps 1 and 2 are best done in small groups with the entire class working at once. The following, more complex steps can be done in small groups with the entire class working at once or by individuals or groups working in a learning center over several days.

1. Challenge each group to assemble examples of as many simple machines as they can from their Legos®. At minimum, they should build a lever, a pulley, a set of gears and a wheel and axle. Encourage students to try several different kinds of levers. (See Figure 1 for sample levers.)

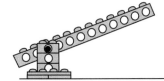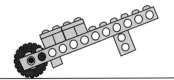

Figure 1: Sample levers built with Legos®

2. If students have not had hands-on experience with levers, they should test the lifting ability of their levers by putting weights such as pennies or paper clips on the load end and then stacking the same kind of weights on the other end until the load is lifted. They could also investigate the effect of changing the position of the fulcrum.

3. Discuss why simple machines are used. (See the "Explanation.")

4. Tell students that they will be designing a special model using Legos®. Explain that this object should demonstrate as many simple machines as possible, all in the same model. The simple machines can be the same or different types. For several days, provide opportunities for individual students, or teams of two or three (but no more) to go to a table where the Legos® are kept and work on this idea. They should make notes on and sketch any ideas they come up with.

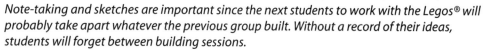

Note-taking and sketches are important since the next students to work with the Legos® will probably take apart whatever the previous group built. Without a record of their ideas, students will forget between building sessions.

5. At the end of the designated time, instruct students to turn in their drawings and notes. Remind students that the drawings will have to be clear and the descriptions easy to understand. Post these and give students time to examine them. Then have designers make a brief presentation. At the end of this process, instruct the class to decide by vote which idea to use. Have the designers build their plan into reality and set it out for display, along with a poster that lists all the simple machines in it.

Explanation

The following explanation is intended for the teacher's information. Modify the explanation for students as required.

Contrary to statements found in many textbooks, simple machines do not allow us to do less work. If we were able to accomplish the same results with less work, we would be able to output more energy than we put in. No simple machine allows us to do this. In fact, due to friction between moving parts, when using simple machines we usually put in more work than we get out in results.

Why, then, do we use simple machines? In most cases, they allow us to accomplish the same result (lifting a heavy box, for instance) using a smaller force that is easier for our muscles to generate. However, this advantage requires a trade-off. If less force is used, we must move the force over a greater distance. Why? Because the work done (the product of force and distance) stays the same, in order to use a smaller force, we must simultaneously increase distance. For example, imagine lifting a 100-newton box up into the back of a truck 1 m above the ground. We can lift it straight up using a 100-newton lifting force and doing 100 joules of work or we could slide it up a 4-m-long inclined plane using a 25 newton force and still doing 100 joules of work. If we were really moving the box up the inclined plane, we would actually need a force somewhat larger than 25 newtons because of the friction between the inclined plane and the box. The same principle applies to doing work with other simple machines such as levers, gears, and multiple pulley systems. The applied force required is smaller, but must move a greater distance.

Another reason simple machines are used is to make something work faster than we can easily do by applying a force directly to it. An example of this use is a hand-operated drill. The bit must turn much faster than you can easily move your hand. By using a progression of larger to smaller gears, the bit can be made to turn much faster than your hand is moving. The trade-off is that your hand must move in a much larger circle (travel a greater distance) than if you were turning the bit directly with your hand.

Extension

Put uniform masses (for example, bricks) on one end of a seesaw. Have the students experimentally determine how many bricks must be stacked at different locations on the other side in order to lift the load.

Cross-Curricular Integration

Art:

- The drawings for the designs will need to be clear and follow rules of proper proportion and perspective. This can be a good lesson in art as a form of accurate communication.

Language arts:

- The written descriptions of the design plans will need to be clear, coherent, and appropriately organized. Using peer editors to proofread will be especially effective here.
- Read aloud or suggest that students read the following book:
  - "The Big Parade" in *Einstein Anderson Goes to Bat,* by Seymour Simon (Puffin, ISBN 0-14-032303-1)
    Einstein uses his understanding of levers to trick the school bullies into carrying more than their share of a heavy load.

Further Reading

Cooper, C.; Osman, T. *How Everyday Things Work;* Facts on File: New York, 1984. (Teachers)

Taylor, B. *Get It in Gear: the Science of Movement;* Random House: New York, 1991. (Students)

Walpole, B. *Fun With Science: Movement;* Warwick: New York, 1987. (Students)

Contributor

Anita Kroger, Gifted and Talented Specialist, Cincinnati, OH; Teaching Science with TOYS, 1986–87.

GEAR UP WITH A LEGO® HELI-TRACTOR

Students explore the transmission of energy using gears.

Lego® Technic gears

GRADE LEVEL

Science activity appropriate for grades 4–8
Cross-Curricular Integration intended for grades 4–6

KEY SCIENCE TOPICS

- simple machines
- energy
- transformation of energy
- work
- inventions

STUDENT BACKGROUND

Students should be familiar with the basic types of simple machines.

KEY PROCESS SKILLS

- communicating   Students explain orally or in writing the function of the gears in their Heli-Tractor.

- classifying   Students build a Lego® Heli-Tractor model and use the model to classify simple machines.

TIME REQUIRED

Setup	10	minutes
Performance	60	minutes*
Cleanup	10	minutes

*Two 30-minute classes. The time will vary depending on the amount of prior experience with Legos®.

Materials

Per group:
- Lego® Technic #8024

Most Lego® Technic sets will have at least one model in the instruction booklet that would be suitable for the activity. If a model other than #8024 is used, the instructions will need to be revised as appropriate.

- zipper-type plastic bag for storing parts

Per class:
- (optional) Lego® Technic #8024 for building sample

Safety and Disposal

No special safety or disposal procedures are required.

Getting Ready

You may want to build a sample heli-tractor before class. Write the steps of the "Procedure" on index cards (one set per group of students).

Part A is independent of Part B and may be done as a stand-alone activity. The model in Part B is more difficult to assemble and may not be appropriate for the younger end of the grade-level range.

Procedure

Part A: Changing Speed

Have students do the following:

1. Before you begin construction, count the number of teeth on the small gear and the large gear.

2. Begin construction by completing the first five steps on pages 8–9 of the instruction booklet that comes with the Lego® set.

3. List the simple machines you can identify in the model at this stage.

4. Do step 6 in the construction instructions. Turn the rear axle (without the wheel) and watch the gears turn. Do the two vertical gears turn in the same direction or opposite directions? Explain why they work this way.

5. Can you tell which gear turns faster? Use a washable marker to mark one of the teeth on each gear. Slowly turn the rear axle and count the number of turns the small axle makes when the large one turns once. How could you have figured this out ahead of time? Can you devise a general rule that would work for all gears?

6. Do step 7 in the construction instructions. Push the tractor along the table. All the gears should turn. Identify what causes each gear to turn.

7. Do steps 8 and 9 in the construction instructions. Are there any levers in your model? If so, where are they?

8. Complete the heli-tractor. Look carefully at all the gears. One of the gears does not really function as a gear. Which one is it? What is its purpose? How do you know it is not really used as a gear?

9. After students have completed Steps 1–8, challenge each group to make a change in the design. Each group should then explain either orally or in writing what their change was and how it works.

Two groups may combine to have a larger number of parts available.

Class Discussion

Calling on each group in turn, ask students to report their answers to the "Procedure" questions. Discuss why the general rule devised in Part A, Step 5, of the "Procedure" must be true.

Part B: Changing Force

Have students do the following:

1. Build the model on pages 13–19 of the instruction booklet.

2. Turn the large gear to raise and lower the forklift. Describe in your own words the chain of events that causes the lift to move.

3. Do the two gears move with the same speed?

4. Do you know why the gear that moves the rack up and down is smaller than the gear you turn? Think about the gears on a ten-speed bike. Why are the gears on the rear wheel used to go uphill larger than the ones used on level ground?

Class Discussion

Have someone describe the sequence of forces that lift the forklift. Solicit ideas for the answer to the question in Part B, Step 4, of the "Procedure." Ask the students questions to lead them into thinking in terms of force and energy if this does not arise naturally in the discussion. You might want to let someone bring in a bike or take the class out to the bike rack to look at the gears. Have some students describe their experiences with multispeed bikes.

Explanation

The following explanation is intended for the teacher's information. Modify the explanation for students as required.

During this activity, your students are able to observe some properties of gears. A gear is a simple machine. It is a toothed wheel that precisely meshes with another toothed wheel. Two or more gears are used to change the size of the force or the speed of the gear or to change direction of motion.

Gears can be connected to one another in two ways. They can either mesh with one another or be connected by a chain. (See Figure 1.) Two gears that mesh will turn in opposite directions. Two gears that are connected by a chain turn in the same direction.

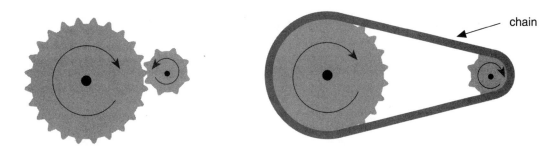

Figure 1: Gears can mesh with one another (left) or be connected by a chain (right).

Gears of different sizes can be used to change speed. A small gear connected to a large gear will turn faster than the large gear. This is because the total distance traveled by a tooth on the small gear must be the same as the distance traveled by a tooth on the large one. If the circumference of one circle is three times bigger than the other, then the smaller gear must turn three times around for every complete turn of the larger one. The 24-toothed gear will turn the eight-toothed gear three times. (See Figure 2.) In general, the ratio of the number of teeth of the two gears is inverse to the ratio of their angular speeds.

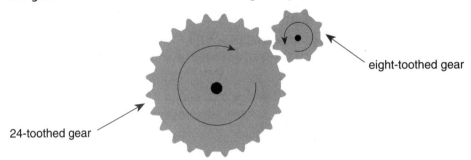

eight-toothed gear

24-toothed gear

Figure 2: The 24-toothed gear will turn the eight-toothed gear three times.

As mentioned previously, one of the functions of a gear is to change the size of a force. This can be accomplished when two gears with different circumferences are mounted on the same shaft. In this case, they turn at the same speed, but the teeth travel different distances and apply different amounts of force. When a force is applied to one gear to make the shaft turn, the other gear can be used to apply a force to something else. Since energy input must equal energy output (ignoring losses to friction), the size of the force can be increased or decreased in proportion to the circumferences of the gears.

Here's how gears change force: Work done equals force times distance. The input energy in one revolution is the applied force times the circumference of the gear to which it is applied. The output energy is the circumference of the output gear multiplied by the force applied by it to something else. If the input gear is larger than the output gear, then the force is increased. (See Figure 3.) If the input gear is smaller than the output gear, the force is decreased. (Force times circumference is the same for each gear.) Just as in changing the speed, the number of teeth can be used to find the ratio of the forces. If the input gear has 24 teeth and the output gear has 12 teeth, then the output force will be twice as large as the input force.

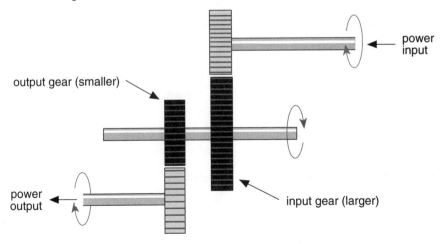

power input

output gear (smaller)

power output

input gear (larger)

Figure 3: If the input gear is larger than the output gear, then the force is increased.

Teaching Physics with TOYS

During the "Procedure," your students are asked to make some specific observations. In Part A, Step 8, they are asked to identify a gear that serves no purpose as a gear. It is on the upper shaft, which turns the rotor blades, and does not mesh with any other gear; thus, it serves no purpose as a gear. It does serve to hold the axle in the proper position for the gears below it to mesh.

For the challenge in Part A, Step 9, of the "Procedure," an easy and interesting change is to use a rubber band to connect the pulleys in the rotor and the tail, which will cause the tail to turn. The tail must be redesigned slightly so the rubber band can move freely.

This model is an excellent example of the rack and pinion system used in the Push-n-Go® if you are also doing that activity. The only difference is that here the pinion gear moves the rack, and in the Push-n-Go® the rack turns the gear.

Cross-Curricular Integration

Social studies:
- Study the invention of labor-saving machines. Students may enjoy reading *Steven Caney's Invention Book* (Workman, ISBN 0894800760).

Further Reading

Cooper, C.; Osman, T. *How Everyday Things Work;* Facts on File: New York, 1984. (Teachers)

Teacher's Guide to Technic; Lego® Systems: Enfield, CT, 1985. (Teachers)

Walpole, B. *Fun With Science: Movement;* Warwick: New York, 1987. (Students)

Contributor

Vivian Schulter, Oak Hills School District, Cincinnati, OH; Teaching Science with TOYS, 1986–87.

LEVITATION USING STATIC ELECTRICITY

Students use a homemade toy to demonstrate the force exerted by static electricity and to demonstrate that like charges repel.

The author levitating a ring using static electricity

GRADE LEVELS

Science activity appropriate for grades 4–12
Cross-Curricular Integration intended for grades 4–6

KEY SCIENCE TOPICS

- electrostatic forces between charged objects
- addition of forces

KEY PROCESS SKILL

- inferring   Students infer the presence of like charges on the rod and ring as they repel each other.

TIME REQUIRED

Setup	25	minutes
Performance	25	minutes
Cleanup	5	minutes

Materials

For "Getting Ready" only

A saw, intended for teacher use only, is needed to make the levitating apparatus the first time the activity is done.

- hacksaw or power saw

For each apparatus

- 40–45-cm section of ½-inch PVC water pipe

The exact length is not important—a 10-foot piece will make 7 sections.

- 1 sheet of ⅜-inch thick foam, as sold for mattress pads, cut into a 25-cm x 40-cm piece and a 10-cm x 15-cm piece

The exact dimensions are not important.

- 1 of the following combinations:
    - half of a plastic soda straw and 1 ball of clay about 1 inch in diameter
    - cork to fit end of pipe and hors d'oeuvres pick, toothpick, or pipe cleaner

- a 2.5-cm x 50-cm strip of ¹/₃₂-inch polyethylene packing material

This material may be found on items that have been wrapped for shipping, or can be ordered from sources that sell packing materials such as Microfoam by Amtek. Another alternative is to find a business that uses the material, and ask them to donate a small quantity.

- hot glue gun and glue sticks
- transparent tape

Safety and Disposal

If any students help with the construction of the apparatus, caution should be used with the glue gun to make sure no one is burned with hot glue. Caution students about not poking one another with the wands. You may want to move to a larger area such as the gym or cafeteria. If you have a small classroom and are unable to move to a larger space, you may want to only allow a few students to use the apparatus at a time. No special disposal procedures are required.

Getting Ready

The apparatus used in this activity was originally sold as a game called "Mystic," by Knots, Inc., but is no longer available. By gathering some common materials, you can make a suitable homemade version. Assuming you make several, it should be possible to gather all materials at a cost of less than $1 per apparatus.

For this activity, you will need to make the following items (See Figure 1):

- plastic rod with tip,
- foam slide,
- mat, and
- ring.

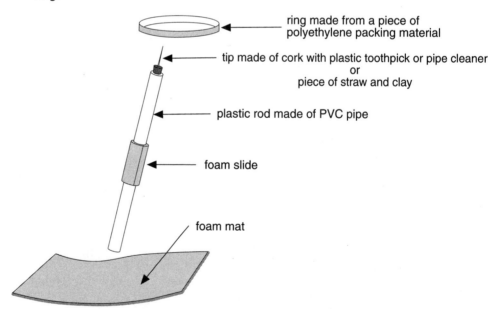

ring made from a piece of polyethylene packing material

tip made of cork with plastic toothpick or pipe cleaner
or
piece of straw and clay

plastic rod made of PVC pipe

foam slide

foam mat

Figure 1: Make a levitation apparatus.

1. To make the plastic rod, cut the PVC pipe to a length of 40–45 cm using a hacksaw or power saw.

Not all PVC pipe is identical. Some pipe allows much better sliding. It is wise to take some foam along to the store and try it out to see if the foam slides up and down the pipe smoothly.

2. The pipe needs a tip in order to pick up the charged ring, because it will cling to you if you touch it with your hands. A cork with a toothpick in it will work but may be considered too dangerous when children are involved. Clay and a section of drinking straw is a safe substitute, but the clay can sometimes make a mess when rolled over the foam mat (as described in Step 4c of the "Procedure"). Another safe alternative is to bend half of a pipe cleaner into the shape of a lollipop and stick the stem into a cork.

3. To make the foam slide, wrap the smaller piece of ⅜-inch-thick foam padding around the rod and put a bead of hot glue on the overlap line. Hold until cool, then cut away the excess. Be careful to leave it loose enough to slide easily.

4. The larger piece of foam padding will be the mat.

5. To make the ring, tape the ends of the packing material strip together with transparent tape. If you wish to use other shapes of packing material besides the ring, go ahead—be imaginative, but remember that the more massive the flying object, the more difficult it will be to levitate. Some reduced patterns are provided at the end of this activity.

 Sometimes polyethylene packing material is treated to be "antistatic." This is particularly likely if the material came wrapped around a stereo or computer. This treated foam will not work for the experiment.

Introducing the Activity

Review the concept of static electricity. Have students recall demonstrations of static electricity they've seen. Explain that you will be using a homemade toy to demonstrate static electricity and that games can be played with this toy.

Procedure

1. Show and demonstrate the sample apparatus. (See Step 4.)

2. Tell the students you will give them a chance to use the toy shortly, and that you want them to think about the different forces that are involved as the toy works. Tell them they will work in pairs, and that they should discuss these forces with their partners as they experiment.

3. Make sure the students know the names for each part of the apparatus. This is important because with three different cylindrical parts, it can get confusing. The rod is the 45-cm piece of PVC pipe. The end of the rod that has the cork or clay in it is top of the rod. The foam cylinder around the rod is the slide. The loop of packing material is the ring. The flat piece of foam is the mat.

 You may want to distribute copies of Figure 1 to the students.

4. Hand out written instructions or put them on the board. Then have the students read the instructions as you demonstrate them, as follows:

   a. Lay the ring flat on the mat.

   b. Move the slide back and forth along the rod about a dozen times using a complete stroke from one end of the rod to the other. (This charges the rod.) Finish with the slide at the top of the rod.

c. Keeping the slide near the top of the rod, roll the slide several times across the ring as it lays on the mat. Turn the ring over and roll the slide over it several more times. (This gives the ring the same charge as the rod.)

d. Once again, move the slide back and forth along the rod several times. This time stop with the slide at the bottom of the rod.

e. Do not touch the ring with your hand; it will cling to you if you do. Using the tip of the straw or pick at the top of the rod, pick up the ring and toss it into the air. Move the top of the rod underneath the ring. The ring will stay up in the air.

5. Show the students any other shapes for the levitated object that you have available. Explain that those will work also, but that you may notice some differences in how they work, such as the height at which they levitate.

6. Have the students try out their apparatus. They should take turns with their partners, discuss the forces involved, and use the alternate shapes if available.

7. Have the students share the ideas they discussed with their partners. Discuss the forces involved, clarifying any misconceptions. (See the "Explanation.")

Explanation

 The following explanation is intended for the teacher's information. Modify the explanation for students as required. In particular, it is not necessary to discuss atoms and electrons with elementary students. You can just discuss moving charges from one object to another.

In this activity, the ring floats because an upward electrostatic force (static electricity) is equal to the downward force of gravity on the ring.

Where does the electrostatic force come from? When the slide is moved up and down the rod or rolled across the ring, some atoms of the foam lose electrons, giving the foam a positive charge. The electrons from the foam are transferred to the rod and the ring, giving both excess negative charges. (No charge is created; it merely is moved from one object to another.)

Now the rod and the ring have the same charge, and the general rule is that like charges repel. Therefore the ring is pushed upward, away from the rod. The electrostatic force gets smaller as the ring and the rod get further apart. The ring will float at a distance above the rod so that the upward electrical force and downward gravitational force are equal. Because the electrostatic force is not large, levitation is possible only when the weight of the ring is small. Heavier shapes float lower because a larger electrostatic force is needed to balance their weight.

Most static electricity experiments do not work well on humid days because a very thin layer of water forms on the surfaces of most objects. Since water is a good conductor of electrons, the electrons can easily move away from the area where you want them to be. The levitation apparatus will work even on humid days, but you will not be able to build up as large an electric charge, so the distance between the rod and the ring will be smaller.

Variations

Hold a contest to see who can keep their ring in the air the longest. Have a relay race to see which team can pass the ring along without letting it drop to the floor.

Extensions for Secondary Students

Secondary students may perform a quantitative investigation of the relationship between the amount of charging and the height of the object above the rod at equilibrium. Have the students measure the mass of the levitated ring and the equilibrium distance. (Either a balance appropriate for measuring a very small mass or a large quantity of ring material will be needed.) Then have them estimate the number of electrons required to produce levitation.

Let's do a sample calculation. The rod in the photo on the first page of this activity is 60 cm long. Using this information, the height of the ring above the top of the rod can be estimated to be 20 cm. For simplicity, you can assume that all the charge is at the tip of the rod. (Because of this assumption, the calculation will result in a number of electrons lower than actually required, since it actually takes a larger number of electrons spread out over the length of the rod to produce the same force.)

The mass of the ring is 4.0×10^{-4} kg. Thus, the weight of the ring is 3.9×10^{-3} newtons. Assuming the amount of charge on the rod and ring to be equal, the students can find the magnitude of the charge as shown:

weight = Coulomb force
$mg = kQ^2/d^2$
3.9×10^{-3} N = $(9.0 \times 10^9 \text{ Nm}^2/\text{C}^2)(Q^2/(0.20\text{m})^2)$
1.3×10^{-7} C = Q

Using the fact that the charge on one electron is 1.6×10^{-19} C, the students should find that the production of this charge requires 8.1×10^{11} (about 800 billion) excess electrons.

Students may find it interesting to compare the number of excess electrons needed to the number of total electrons present in the ring. The empirical formula for ethylene is C_2H_4, so one molecule has a mass of approximately 4.7×10^{-26} kg. Thus, there are 5.2×10^{22} atoms in the band [$(4.0 \times 10^{-4}$ kg/4.7×10^{-26} kg) $\times$ 6]. Although 8.1×10^{11} electrons may sound like a very large number, only a tiny fraction of the atoms in the ring have lost or gained electrons.

Cross-Curricular Integration

Language arts:
- Students may wish to cut out a shape that represents a creature of some kind, or perhaps a spaceship (patterns provided). They may write a highly imaginative story about groups or colonies of creatures that depend on the presence of electrostatic forces to allow them to fly or travel above the surface of the Earth. They might even write a play about such creatures and act it out using the toy.

- Read aloud or suggest that students read one or more of the following Einstein Anderson stories by Seymour Simon:
  - "The Electric Spark" in *Einstein Anderson Shocks His Friends* (Puffin, ISBN 0-670-29070-x)
    Einstein shocks someone by rubbing his arm against his nylon jacket to build up an electric charge and then touching the person's nose.
  - "The Spring Festival" in *Einstein Anderson Lights Up the Sky* (Puffin, ISBN 0-670-29066-1)
    Some students use static electricity to hang the party balloons from the wall and ceiling. Unfortunately, the next day is very damp and all the balloons fall down.

Math:
- Higher math applications are essential for the Coulomb's law investigation described in the "Extension."
- Younger students can record observations of how long or how high different rings can be levitated and compare this to the size of the ring. These results can be graphed or simply compiled in a data chart.

Further Reading

Griffith, W.T. *The Physics of Everyday Phenomena;* Wm. C. Brown: Dubuque, IA, 1992, pp 216–224. (Teachers or older students)

Kirkpatrick, L.; Wheeler, G. *Physics: A World View;* Saunders College: Philadelphia, 1992, pp 420–434. (Teachers or older students)

White, J.R. *The Hidden World of Forces;* Dodd, Mead: New York, 1987, pp 99–104. (Teachers or students)

Handout Master

A master for the following handout is provided:
- Patterns for Flying Objects

Copy as needed for classroom use.

LEVITATION USING STATIC ELECTRICITY

Patterns for Flying Objects*

* These patterns are reduced 75%.

DOC SHOCK

Use the Operation® game to explore the structure and components of an electric circuit.

Operation® game

GRADE LEVELS

Science activity appropriate for grades 4–6
Cross-Curricular Integration intended for grades 4–6

KEY SCIENCE TOPICS

- conductors and nonconductors or insulators
- electricity
- energy transfer
- open and closed circuits
- parts of a circuit

STUDENT BACKGROUND

Before doing the activity, the students should be introduced to the idea of a complete circuit via some experimentation with batteries and bulbs. "Doc Shock" is then appropriate as an application activity on electric circuits.

KEY PROCESS SKILLS

• hypothesizing	Students form hypotheses about why the game board of Operation® lights up and buzzes as they play.
• investigating	Students investigate the inner workings of the game board.

TIME REQUIRED

Setup	10	minutes
Performance	45	minutes
Cleanup	5	minutes

Materials

For "Getting Ready" only
- knife

For the "Procedure"

Per group of 3–4 students

- 1 Operation® game by Milton Bradley Company

This game is available at toy stores, and many students have this game at home.

For "Variations and Extensions"

❶ Per class
- Operation® game
- assorted small objects to test for conductivity such as pencils, paper clips, pennies, marbles, erasers, and plastic construction blocks

❷ Per class
- Operation® game
- a piece of wire about 10 cm long

❸ Per class
- other games or toys in which bulbs light or buzzers buzz

❹ Per group
- paper clips
- cardboard
- insulated wires
- aluminum foil
- batteries
- bulbs

❺ Per group
- buzzer or bulb
- switch
- light bulb
- battery
- wire
- lidded container

Safety and Disposal

No special safety or disposal procedures are required.

Getting Ready

The cover of the Operation® game is held down by plastic pegs. Use a knife to cut the cardboard around the pegs. If students are bringing the game from home, check with parents for permission to remove cover. The cover can easily be replaced, restoring the game to full usefulness.

Introducing the Activity

Explain the game and its rules or allow students who have played it before to explain the game to the class.

Procedure

1. Explain to the students that they will be playing the Operation® game in their small groups. Ask them to think about why the game board is making noise and lighting up as they play.

2. After they play, and before proceeding with the lesson, have the students stop and share with a partner their thoughts about the reasons behind the noise and light or have them write their ideas down.

   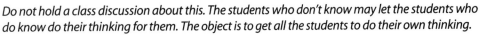 *Do not hold a class discussion about this. The students who don't know may let the students who do know do their thinking for them. The object is to get all the students to do their own thinking.*

3. Have each group of students take the cover off their game board and investigate the internal parts. Ask them to make the bulb light and buzzer buzz, while noticing all the internal parts needed to do this. Have each group list these parts.

4. Have the students share their observations about the necessary components of the game in a class discussion and list the common observations on the board. Match these with general terms that apply to any circuit, as shown below.

   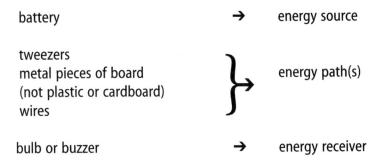

battery	→	energy source
tweezers metal pieces of board (not plastic or cardboard) wires	→	energy path(s)
bulb or buzzer	→	energy receiver

5. Summarize the activity as follows: "All of these components are necessary for an electric circuit to work. If they are all connected and working, the circuit is closed or completed and electric current moves around the entire path. If any connections are not made, the circuit is open and electric current can't move around the entire path. An open circuit is like a drawbridge when it is raised—nothing can move across."

Explanation

➤ *The following explanation is intended for the teacher's information. Modify the explanation for students as required.*

The Operation® game uses a simple electric circuit. Every electric circuit has three essential components: a source of electricity (electrical energy), a receiver or user of electricity, and one or more objects through which the electricity may travel to get from the source to the receiver and back again. The most easily recognized source is the battery. The source could also be a generator, or it could be an AC outlet in the wall through which electricity comes from a generator that might be far away. The receiver could be a bell, a buzzer, a light bulb, a toaster, a curling iron, or an alarm clock. The most common object through which the electricity travels is a covered copper wire, but it could be a metal plate, water, or a human being. An object that works in this capacity—one through which electricity will travel—is called a conductor. An object that will not work in this capacity is called a nonconductor or insulator. There may be additional components in the circuit, the most common of which is probably the switch. A switch is simply a handy device for opening or closing the circuit without disturbing any of the other connections.

When the cover is removed from the game, the circuit's basic components can be seen. The two batteries act as the energy source; the metal plate and the two wires are the path through which the energy flows. The bulb/buzzer is the energy receiver and the tweezers act as a switch, completing the circuit when they touch the metal plate. Figure 1 shows a diagram of this circuit:

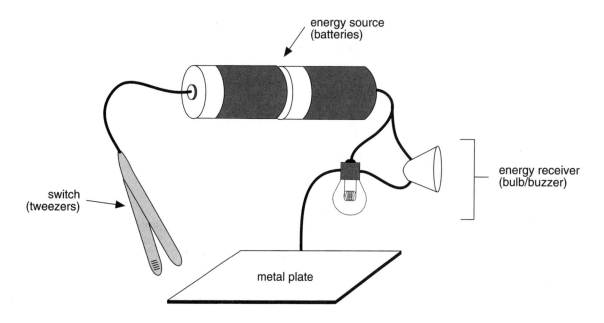

Figure 1: The Operation® game uses a simple electric circuit.

A schematic diagram of this circuit would resemble the one shown in Figure 2.

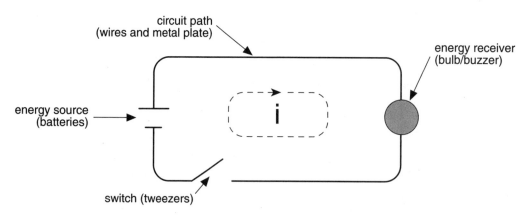

Figure 2: Schematic diagram of the circuit

When the switch is closed (either side of the tweezers touches the metal plate) the circuit path is closed and the current (i) flows.

Extensions

1. While the game board is disassembled, allow students to test other objects to see if they will allow energy transfer. To do this, hold objects with the tweezers and attempt to make a closed circuit by touching the objects to the metal plate. Introduce the terms "conductor" and "insulator."

2. Remove one of the batteries and use a wire to complete the circuit in place of the removed battery. Ask the students to explain the dimmer bulb and slower motor.

3. Have the students bring in other games or toys in which bulbs light or buzzers buzz in order to see how they are constructed.

4. Have students make their own game boards with paper clips, cardboard, insulated wires, aluminum foil, batteries, and bulbs.

5. Have the students read *Dear Mr. Henshaw*, by Beverly Cleary (Avon, ISBN 0380709589). This is an excellent book, but if time doesn't permit reading the entire book, focus on pages 95–102. A circuit similar to the one Leigh built can be constructed either as a group or class activity. The doorbell described may be hard to find, but Radio Shack sells a variety of buzzers that could be used. Also a small light bulb is an excellent (and quieter) substitute that uses the same principles. For this activity each group will need a switch, buzzer or light bulb, battery, and wire, as well as a lidded container. Have the students demonstrate that they understand how and why switches and electric circuits work. After the group project is completed, leave the materials out in a learning center with reminder cards so that students may individually experiment and make their own alarm.

Cross-Curricular Integration

Language arts:
- After completing the activity, have the students write a simple descriptive paragraph about the electric circuit. It will frequently be easier to write a clear main idea, supporting sentences, and closing summary statement about the circuit than about "My kitten" or "My backyard," because in these other topics the students have so many potential ideas to include and have to make so many choices. Later you may wish to have the students write a comparison/contrast paragraph between open and closed circuits or other similar, but not identical, circuits.

Further Reading

Ardley, N. *Discovering Electricity;* Franklin Watts: New York, 1984. (Students)

Faughn, J.; Turk, J.; Turk, A. *Physical Science;* Saunders College: Philadelphia, 1991. (Teachers)

Kirkpatrick, L.; Wheeler, G. *Physics: A World View;* Saunders College: Philadelphia, 1992. (Teachers)

Thier, H.D.; Knolt, R.C. *SCIS III: Scientific Theories;* Delta Education: Hudson, NY, 1993. (Teachers)

Contributors

Holly Rice, Wyoming Middle School, Wyoming, OH; Teaching Science with TOYS, 1988–89.

Hope Spangler, Graduate Student, Physics Department, Miami University, Oxford, OH.

Activities for Grades 7–9

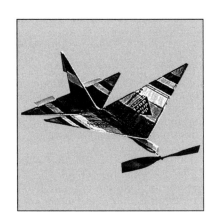

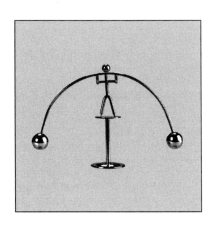

WALKING FEET

Students form an understanding of how time and distance are related to the average velocity and acceleration of an object.

A set of Phantom Feet® in action

GRADE LEVELS

Science activity appropriate for grades 7–9

KEY SCIENCE TOPICS

- acceleration
- kinematics
- motion
- velocity

STUDENT BACKGROUND

Students should know how to use a stopwatch and produce graphs and should be familiar with metric measurement. Before the start of the activity, review the definitions of velocity and acceleration.

KEY PROCESS SKILLS

• collecting data	Students collect time and distance data for the walking toy.
• making graphs	Students construct position-time graphs to investigate velocity and acceleration.

TIME REQUIRED

Setup	5–10	minutes
Performance	20–40	minutes
Cleanup	5–10	minutes

Materials

For the "Procedure"
Per pair of students
- Phantom Feet® by TOMY Corp. or similar wind-up walking toy
- meterstick
- 20–30 colored toothpicks
- stopwatch

 It's not vital to the activity that each pair of students have a stopwatch. For example, if you have one stopwatch, you can call out time intervals for the class.

For the Variation
- duct tape
- stop watch

Safety and Disposal

No special safety or disposal procedures are required.

Procedure

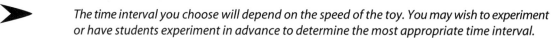

1. Choose a time increment to use during the activity, for example, 1-second or 5-second intervals.

   The time interval you choose will depend on the speed of the toy. You may wish to experiment or have students experiment in advance to determine the most appropriate time interval.

2. Tell students that they will be placing a colored toothpick on or next to the meterstick to mark the position of the walking toy as each time increment is called out.

3. Have each pair of students place the walking toy by the meterstick and mark the starting position.

   Each team should experiment to see if their toy veers from a straight path. If it does, they should put their meterstick on the side to which the toy veers.

4. Tell them to start the walking toy and the stopwatch and mark the positions until the toy stops.

5. Have students work in pairs to collect data. One student will read the positions that were marked at time intervals. The other will record the information in table form.

6. Instruct students to make position-time graphs.

   Walking toys typically maintain a constant velocity for a while and then decelerate to a stop.

7. Have the students calculate average velocities from either their tables or graphs, for example, average velocities for the entire trip, for the time interval from the start until the toy begins to slow down, for the interval of highest velocity, or for each of the recorded time intervals. If the average velocities for each time interval are obtained, these velocities can be used as the velocities at the midpoints of the time intervals to make a velocity-time graph. The velocity-time values can then be used to calculate average accelerations.

8. (optional) Have one team of students put its data, graph, and calculations on the overhead projector or the board for classroom discussion. For example, the highest average velocities obtained by the groups can be compared.

Variation

An interesting way to introduce these kinematic quantities is the Hallway Walk. Pairs of students attempt to walk at a constant velocity down a hallway where marks have been made at equal position intervals. Time is recorded at each mark during the walk. After the walk, the position and time data can be used to calculate average velocities as in the "Procedure." The marks can be made by placing duct tape across the hallway; a useful interval is 5 m. One member of the team carries a stopwatch, maintains a constant velocity, and calls out the time at each mark using the lap timer feature of the stopwatch. The other member walks alongside and records the times as they are called out. The students can be advised to try a slow walk first, and later, after calculations have been done, they can be asked to try to double or halve the previous velocity. The average velocities for each interval and for the entire trip can be calculated. Prizes can be given for the most constant pace over all the intervals and for the best attempt at doubling the velocity.

Explanation

 The following explanation is intended for the teacher's information. Modify the explanation for students as required.

In this activity, students collect time and distance data that enable them to calculate the two quantities defined in the study of motion, or kinematics: average velocity and average acceleration. Velocity is the rate of change of position, and acceleration is the rate of change of velocity. The quantities used in the definition of average velocity are displacement and time interval. Displacement is change in position or difference in final and initial positions. The time interval is the difference in final and initial times. Average velocity is defined as

$$V_{avg} = \frac{x_f - x_i}{t_f - t_i}$$

If position is graphed vertically and time horizontally, the graph is referred to as a position-time graph. According to the definition of average velocity, the slope of the line joining any two points is the average velocity over that time interval. A typical position-time graph for a walking toy is shown in Figure 1. The data points have been connected by straight lines that extend over the 5-second time intervals. In this example, the velocity is constant at 2 cm/s over the first 15 seconds of the trip, and then the toy slows down over the final 10 seconds.

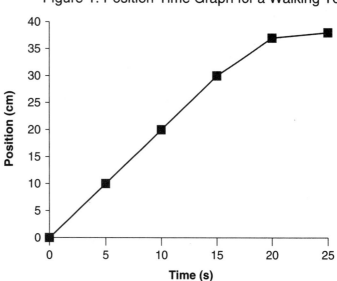

Figure 1: Position-Time Graph for a Walking Toy

The two quantities used in the definition of average acceleration are velocity change, or difference in final and initial velocities, and time interval. Average acceleration is defined as

$$a_{avg} = \frac{v_f - v_i}{t_f - t_i}$$

If velocity is graphed vertically and time horizontally, the graph is known as a velocity-time graph. According to the definition of average acceleration, the slope of the line joining any two points is the average velocity over that time interval. A velocity-time graph for the walking toy in Figure 1 is shown in Figure 2. The velocities are the average velocities for the 5-second time intervals connected by straight lines in Figure 1, and they are plotted at the midpoints of the intervals. For this example, the acceleration is zero for the first part of the trip, since the velocity is constant there. During the last part of the trip, the acceleration is negative, called a deceleration, and it is approximately -0.2 cm/s². Straight lines extending over the 5-second time intervals have been drawn through the data points.

Figure 2: Velocity-Time Graph for a Walking Toy

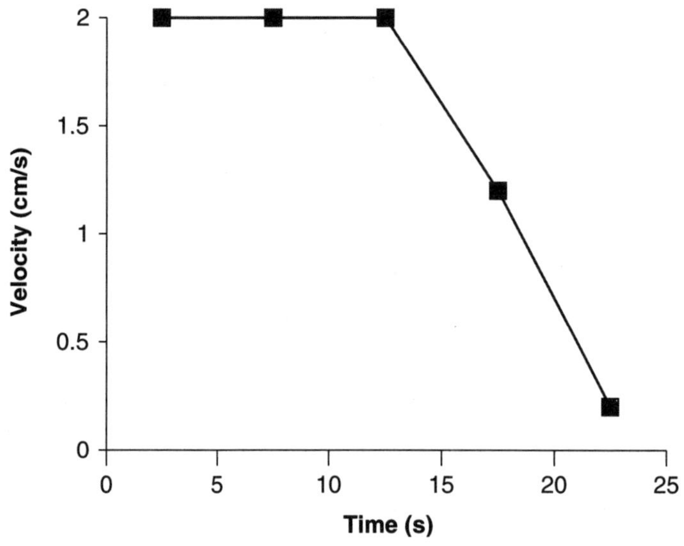

Contributor

Gary Lovely, Edgewood Middle School, Seven Mile, OH; Teaching Science with TOYS peer mentor.

DOWNHILL RACER

Students use an inclined plane to calculate constant acceleration due to gravity.

Hot Wheels®-type car

GRADE LEVELS

Science activity appropriate for grades 7–9
Cross-Curricular Integration intended for grades 7–9

KEY SCIENCE TOPICS

- acceleration and acceleration due to gravity
- displacement
- mean and deviation
- velocity

STUDENT BACKGROUND

Acceleration due to gravity in free fall should be discussed in detail. This activity works well if students have already completed the Teaching Science with TOYS activity "Walking Feet," which introduces distance and velocity graphs.

KEY PROCESS SKILLS

• measuring	Students measure constant acceleration due to gravity using a downhill racing car.
• controlling variables	Students vary the height of the ramp to vary the acceleration.

TIME REQUIRED

Setup	5	minutes
Performance	20	minutes
Cleanup	5	minutes

Materials

For the "Procedure"
Per group
- 1 Hot Wheels®-type car
- 1 board, at least 1.5 m long and thick enough not to bow when inclined
- meterstick
- timer with second hand

> *It's not vital to the activity that each group of students have a timer. For example, if you have only one timer, you can call out time intervals for the class.*

- protractor
- blocks of various heights
- (optional) tape or clamp
- graph paper

For "Variations and Extensions"
Per group
- split-time stopwatch that can record several times, or several stopwatches

Safety and Disposal

No special safety or disposal procedures are required.

Introducing the Activity

To generate interest in this activity, have the students attempt to measure the effect of gravity on an object by timing a short free-fall drop. This will be nearly impossible. Tell the students there is a way to measure the effects of gravity on a falling object by reducing the effect of gravity with the use of an inclined plane or ramp. This will reduce the acceleration to a more measurable value.

Procedure

1. Divide the students into groups and assign jobs.
Each group will need a Car Holder, several Distance Markers, a Measurer, a Recorder, and a Timer to perform the following jobs:
- *Car Holder—to hold the car at the top of the ramp and release it;*
- *Distance Markers—to use a finger to mark the position of the car at a particular time interval;*
- *Measurer—to measure the distances from the zero point to each mark made by the Distance Markers;*
- *Recorder—to record the Measurer's findings; and*
- *Timer—to call out 1-second time intervals.*
The same student can perform more than one job.

2. Assign time intervals to the Distance Markers.

3. Have each group place one end of the board on a block to form a ramp (an inclined plane) and measure the ramp angle with a protractor.
It is best to start with a very small ramp angle (angle between the ramp and the horizontal), such as 5° or 10°, since students will need for the car to stay on the ramp for at least 3 seconds.

4. Use one group's setup to show the class that the car will roll down the ramp more slowly than it would free-fall from the same height.

5. Have the Car Holder in each group hold the car at the top of the ramp. This position is zero.
A meterstick can be taped or clamped along the ramp to facilitate measuring. If a car will not travel in a straight line, metersticks could be taped on each side of the ramp to prevent the car from falling off.

6. Instruct each Car Holder to release the car. As soon as the car is released, have the Timer begin to call out 1-second intervals until the car leaves the ramp.
You may wish to have students continue to collect data until the car stops moving. (The car will start decelerating after it leaves the ramp.)

7. Instruct each Distance Marker to mark the position of the car with a finger when his or her assigned time interval is called out.

8. After the car has left the ramp, have each Measurer measure the distance from the zero point to each mark denoted by the Distance Markers and record this information in a data table.

9. Instruct each group to make a graph of position versus time using the data gathered above.

➤ *See Figure 1 for a sample position-time graph.*

10. Have groups use their data to calculate the average velocity v_{avg} between every pair of points and then make a graph of v_{avg} versus time, plotting v_{avg} at the midpoint of each time interval.

➤ *See Figure 2 for a sample velocity-time graph.*

11. Instruct groups to find the slope of the best-fit line through the data points on the velocity-time graph and compare this with the theoretical value of the acceleration, $a=g \sin \theta$, where θ is the ramp angle and g is acceleration due to gravity. Also, use the measured acceleration to calculate the theoretical value of the displacement x of the car:

$$x = \frac{at^2}{2}$$

and compare this with the actual displacement.

Variations and Extensions

1. Increase the ramp angle and repeat the data collection process. If desired, continue measuring the car's motion until it slows and stops. Your graphs will then show acceleration and deceleration.

2. Repeat the activity several times and use the average values of the measured times or displacements to make the position versus time graph. Use the scatter in the data to introduce the concept of experimental error. A simple method of calculating the error in each data point is to use plus or minus the deviation from the mean of the data point that is furthest from the mean. For example, if repeated measurements of the time at the 1-m displacement are 1.3, 1.3, 1.4, 1.4, and 1.6 seconds, the average time is 1.4 seconds and the error is plus or minus 0.2 seconds. (1.6–1.4=0.2 seconds.)

3. Instead of measuring the distance traveled in specified time intervals, measure the time required to travel specified distances. Mark regular distance intervals such as 0.5 m on the ramp. Use a stopwatch or several stopwatches to record the time at which the car passed each mark.

Explanation

➤ *The following explanation is intended for the teacher's information. Modify the explanation for students as required.*

In this activity, students collect time and distance data for a situation in which the velocity is not constant and there is acceleration. They use these data to calculate average velocity and average acceleration. A ramp is used as a convenient way to observe accelerations considerably smaller than the acceleration due to gravity. This is because the component of acceleration due to gravity in the down-the-plane direction is $g \sin \theta$ where g is the acceleration due to gravity and θ is the angle of incline of the ramp.

When you let go of the car at the top of the ramp, it begins to roll forward. As it continues down the hill, its position changes more quickly. The car covers a greater and greater distance in each equal time interval. The change in position over time is called velocity. Average velocity is determined by displacement divided by time interval, where displacement is change in position or the difference in final (f) and initial (i) positions:

$$v_{avg} = \frac{x_f - x_i}{t_f - t_i}$$

A position-time graph is a plot of position vertically versus time horizontally. The slope of the line joining any two points is the average velocity over that time interval. A sample position-time graph is shown in Figure 1.

Figure 1: Sample Position-Time Graph for a Toy Car on a Ramp

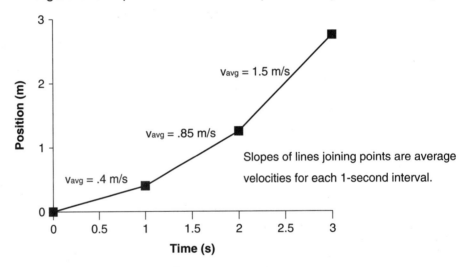

The points do not lie in a straight line; the slopes of the lines joining points are not the same. Thus we know that the velocity is not constant. In our example, the car's average velocity was 0.4 m/s during the first second. By the third second its average velocity was 1.5 m/s. The change of velocity of an object is called its acceleration. Average acceleration is defined as the rate of change of velocity and is determined by velocity change, or difference in final (f) and initial (i) velocities divided by the time interval:

$$a_{avg} = \frac{v_f - v_i}{t_f - t_i}$$

You can plot velocity versus time much as you previously plotted distance versus time. An example of this is shown in Figure 2. The velocities are average velocities for the 1-second time intervals connected by straight lines in Figure 1, and they are plotted at the midpoints of the intervals. This time, the slope of the line connecting any two points is the average acceleration over that time interval. Theoretically, in an inclined plane experiment like this one, the average acceleration (slope of the line) over each time interval should be the same: acceleration should be constant and the data points should lie in a straight line with an upward slope. However, in reality, due to tiny irregularities in the ramp surface, the construction of the toy car, and experimental uncertainty, you are very unlikely to get data points that lie exactly in a straight line. As shown in Figure 2, the best-fit line through your points approximates the acceleration of the car.

Figure 2: Sample Velocity-Time Graph for a Toy Car on a Ramp

Slopes of lines joining points (dashed lines) are average accelerations for each interval. These slopes are nearly the same, indicating a nearly constant acceleration.

The slope of the best fit line through all the points (solid line) is the average acceleration for the car's trip down the ramp.

Cross-Curricular Integration

Language arts:
- Have students research and write reports on the acceleration of various objects, such as race cars, family cars, space shuttles, etc.

Physical education:
- Have several students with stopwatches position themselves at 1-m intervals. From a standing start, have one student volunteer race past them. Have the students make a data table of times and distances and calculate the acceleration. Then hold a class discussion about the importance of acceleration to different sports. Is acceleration more important to a sprinter or a long- distance runner?

Social studies:
- Galileo's experiments showed that falling objects of different masses accelerate approximately equally and disproved Aristotle's theory that an object that weighs 10 times more than another should fall 10 times faster than the lighter object. Have students study the lives and work of Galileo and Aristotle.

References

Trinklein, F.E. *Modern Physics;* Holt, Rinehart and Winston: New York, 1992.

Tropp, H.E.; Friedi, A.E. *Modern Physical Science;* Holt, Rinehart and Winston: New York, 1991.

Contributor

Gary Lovely, Edgewood Middle School, Seven Mile, OH; Teaching Science with TOYS peer mentor.

THE PROJECTILE CAR

Students gain an understanding of the conservation of linear momentum.

Projectile Car and BB Corral

GRADE LEVELS

Science activity appropriate for grades 7–9
Cross-Curricular Integration intended for grades 7–9

KEY SCIENCE TOPICS

- force
- impulse
- momentum
- Newton's second law

STUDENT BACKGROUND

Students should have some knowledge of force, Newton's laws, and friction.

KEY PROCESS SKILLS

- measuring — Students measure the distance Projectile Cars move.

- controlling variables — Students investigate changes in linear momentum as the mass of the Projectile Car is varied.

TIME REQUIRED

Setup   10 minutes (plus 1 hour first time)
Performance 30 minutes
Cleanup   5 minutes

Materials

For "Getting Ready" only

These materials, intended for teacher use only, are needed to make the Projectile Car and BB Corral the first time the activity is done.

- drill and bit to start holes for screws (⅛-inch drill bit)
- glue
- screwdriver

For the "Procedure"

Per class or group

- Projectile Car made from the following:
  - block of smooth pine wood, approximately ¾ inch thick, 3½ inches wide, and 8 inches long
  - 2 #10 or #12 wood screws, 1¼ inches long
  - 1 strong rubber band (such as #62)
  - spool of string
  - #6 eye screw, ½ inch long
- BB Corral made from the following:
  - 1 piece of wood (pressed-wood shelving) approximately ¾ inch x 8 inches x 4 feet
  - wood lattice strip (approximately ¼ inch thick x 1⅝ inches wide), 2 pieces 4 feet long and 2 pieces approximately 9 inches long
  - small #17 wire nails approximately 1 inch long
  - wood glue

➤ *This activity can be done as a class project or in groups of two or more students. If done in small groups, each group will require a separate Projectile Car and BB Corral. The cars must have the same mass projectiles and rubber bands if a class comparison is desired. If not, variable strength rubber bands and different masses may be used.*

- several strong rubber bands
- string
- matches or propane starter
- meterstick
- projectiles of different masses

➤ *Standard slotted 100-g and 200-g lab masses are convenient.*

- box of 2,500 or more BBs

For "Variations and Extensions"

❶ All materials listed for the "Procedure" plus the following:
- rubber bands of different strengths

❷ Per class
- water and compressed air toy rocket

Safety and Disposal

No special safety and disposal procedures are required.

Getting Ready

Construct the Projectile Car(s) and the BB Corral(s) as directed below. After the car and BB Corral are complete, set up the BB Corral and test projectiles of different masses and rubber bands of different strengths for desired effect.

Construct a Projectile Car

1. Starting at one end, mark off 1-cm lines across the width of the wood block. (See Figure 1.)

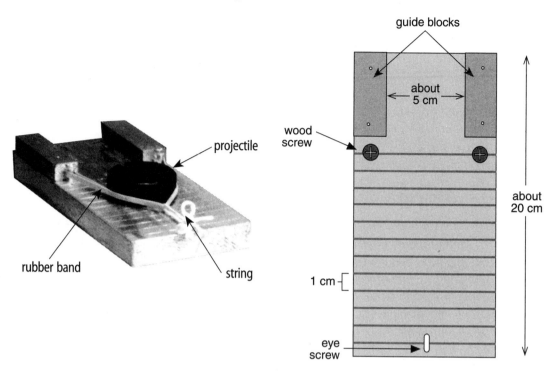

Figure 1: Construct a Projectile Car.

2. At the center line, 1 cm in from each side, drill a pilot hole for each wood screw.

3. Screw the wood screws into the holes so that they do not come through the other side.

4. Screw in the eye screw at the end where the centimeter increments start.

5. Nail the guide blocks to the board as shown in Figure 1.

 They are positioned so they do not interfere with the rubber band but are in a position to guide the projectile. If standard slotted weights are being used, separate the blocks by slightly more than the diameter of the weights.

Construct a BB Corral

1. Cut the pressed-wood shelving material that will form the BB Corral to the desired size.

2. Cut two pieces of the wood lattice strip to the exact length of the BB Corral.

3. Glue and nail the strips to the BB Corral, keeping the strips flush at the ends and along one long side. The strips should form a trough or corral.

4. Cut two pieces of the lattice strip to box off the ends of the corral. Measure the total width of the corral plus the side strips to find the length of the end pieces.

5. Glue and nail these strips to form the ends of the corral.

6. Fill the BB Corral with BBs one layer deep.

Procedure

Have each group do the following:

1. Loop the rubber band over the two wood screws.

2. Cut a piece of string from the spool and tie one end to the center of the rubber band on the Projectile Car.

3. Pull the string until the rubber band is pulled back to one of the centimeter marks and tie the string to the eye screw. Record the mark to which the rubber band is stretched.

4. Place the car on the BBs in the BB Corral.

5. Load the mass to be fired against the rubber band and between the guide strips.

6. Mark the position of the car.

7. Burn through the string with a match or propane starter. (This will release the rubber band without any outside movement. The mass will be fired out, forcing the car to move in the opposite direction.)

8. Record distances of car movement for different masses and different stretches of rubber band.

9. Plot graphs of distance versus mass and distance versus stretch.

10. Discuss with the class how these graphs are related to conservation of linear momentum.

Variations and Extensions

1. Repeat the "Procedure" using different-strength rubber bands.

2. Demonstrate a water and compressed air toy rocket to lead into a discussion of systems with varying mass and the operation of rockets.

Explanation

 The following explanation is intended for the teacher's information. Modify the explanation for students as required.

This activity enables students to investigate Newton's second law, which describes the relationship between force, mass, and acceleration. The law can be stated from two points of view: a force viewpoint (which defines the motion of the object as being caused by a net force on the object) and a momentum viewpoint (which defines the motion of the object as being caused by an impulse). Each of these viewpoints is discussed in turn.

Newton's second law can be expressed in terms of the momentum, $p = mv$, as:

$$F = \frac{\Delta p}{\Delta t} = \frac{\Delta(mv)}{\Delta t}$$

where F is the average net force acting on an object during a time interval Δt, and Δp is the change in the linear momentum of the object during this time interval.

If the mass of the object is constant, the familiar $F = ma$ form of Newton's second law can be obtained (since a change in velocity Δv during a time interval Δt is equal to acceleration a):

$$F = m\frac{\Delta v}{\Delta t} = ma$$

From a force viewpoint, the motion of an object is caused by a net force on the object.

The momentum form of Newton's second law can be written as:

$$F\Delta t = \Delta p$$

where the product $F\Delta t$ is known as the impulse. From a momentum viewpoint, the cause of the motion of an object is an impulse, or a net force acting during a time interval resulting in a change in the linear momentum of the object.

If the net force is zero, the impulse is zero, the change in linear momentum is zero, and the linear momentum is constant. This is known as the law of conservation of linear momentum. When a quantity like the linear momentum is constant, it is said to be conserved.

In the Projectile Car, the projectile and the car can be considered to be two parts of the same system. They exert equal and opposite forces on each other according to Newton's third law. Other horizontal forces acting on the system are friction between the car and the surface on which it moves and the air resistance acting on the projectile and car. Assuming that the friction and air resistance forces are negligible compared with the forces of the car and projectile on each other, the net force is zero and the linear momentum is conserved.

Before the projectile is fired, the linear momentum p_1 of the car and projectile is zero:

$$p_1 = 0$$

After firing, the linear momentum p_2 is given by

$$p_2 - Mv_2 + m(v_2 - v_0)$$

where M = mass of car;
$\quad m$ = mass of projectile;
$\quad v_2$ = velocity of car after firing; and
$\quad v_0$ = velocity of projectile relative to car.

Since the linear momentum is conserved, $p_1 = p_2$:

$$Mv_2 + mv_2 - mv_0 = 0$$

Solve for the velocity v_2 of the car after firing:

$$v_2 = \frac{mv_0}{M + m}$$

The car accelerates from rest to a velocity v_2. From a conservation of momentum viewpoint, the car must accelerate in order to compensate for the momentum carried off by the projectile in the opposite direction. From the viewpoint of Newton's third law, the car exerts a force on the projectile and the projectile exerts an equal and opposite force on the car. This force accelerates the car to a velocity v_2.

Even though we attempt to make the friction force as small as possible by using BBs, the friction force does decelerate the car and eventually bring it to rest after some displacement. The displacement is proportional to the velocity v_2 and therefore increases as the velocity and mass of the projectile increase. The stretch of the rubber band determines the projectile velocity. These relationships can be verified by the students using the graphs they draw from their data.

Cross-Curricular Integration

Physical education:
- Discuss how athletes' masses affect performance in different sports. (Be sure to discuss some contact and non-contact sports.) Why are football players usually bigger than baseball players? Why are linemen usually bigger than running backs? Why is it advantageous for sumo wrestlers to be large?

References

Lamb, W.G.; Cuevas, M.M.; Lehrman, R.L. *Physical Science;* Harcourt Brace Jovanovich: Orlando, FL, 1989.

Trinklein, F.E. *Modern Physics;* Holt, Rinehart and Winston: New York, 1992.

Contributor

Gary Lovely, Edgewood Middle School, Seven Mile, OH; Teaching Science with TOYS peer mentor.

BALANCE TOYS AND CENTER OF GRAVITY

*Introduce the concepts of center of gravity
and stable and unstable rotational equilibrium with fun toys that grab students' attention.*

A wire-frame balance toy

Materials

For "Getting Ready" only
These materials, intended for teacher use only, are needed to make the homemade figures.

- drill or hole punch
- (optional) drawings or photographs of athletes or other figures in motion, no bigger than 15 cm x 15 cm

For the "Procedure"
Part A, per group
- 1 balance toy (such as a small wire-frame figure that does not fall when pushed, but instead rocks back and forth and returns to its original position)
- flexible toy figure or doll
- 6 squares of cardboard or $\frac{1}{16}$-in or $\frac{1}{8}$-in thick plastic, approximately 15 cm x 15 cm
- clothespins without springs
- string

- paper clip
- masking tape
- marker or pencil
- straightedge for geometric figures
- small object such as a nut or bolt
- dime
- belt

For "Variations and Extensions"

❶ Per group
- potatoes
- wire coat hangers
- corks
- rubber stoppers
- Styrofoam™ balls

❷ All materials listed for Part A plus the following:
Per group
- compass

❸ Per group
- wooden or cardboard school box or a plastic videotape case
- a small heavy item, such as a slotted weight, stone, or battery
- (optional) drill

❹ Per class
- a bar or rope for a person to hang from

❺ Per student
- piece of cardstock
- 2 coins
- tape or glue

Safety and Disposal

Use caution in handling any sharp toys or pieces of wire coat hanger (Extension 1). Be sure students do not attempt to balance heavy or otherwise dangerous objects. No special disposal procedures are required.

Getting Ready

Each group will need all of the types of shapes shown in Table 1. Cut out the shapes from the squares of cardboard or plastic. Examples of symmetric figures are circles, bilateral or equilateral triangles, and rectangles. An easy way to make asymmetric figures is to cut randomly around the edges of the square. The body shapes can be in the form of athletes from a variety of sports in both symmetric and asymmetric positions. These human figures will be interesting to the students, but do not need to be detailed artworks. The easiest way to make these human shapes is to do a simplified tracing of the body from a drawing or photograph. Drill or punch several holes around the edge of each of the figures to use as support points.

Table 1: Figures Needed for Each Group			
Description	Example	Description	Example
symmetric, geometric, solid		asymmetric letter	
symmetric, geometric, hollow		symmetric human	
asymmetric, abstract, solid		asymmetric human	

Procedure

Part A: Locating the Center of Gravity

Have small groups of students perform the following steps:

1. Hang a solid symmetric figure from one of its support points and use a string hanging from the support point as a guide to mark a vertical line on the figures with a marker or pencil. A support can be made by opening one end of a paper clip and attaching a string with a small object (such as a nut or bolt) tied to the other end.

2. Hang the same figure from a different support point and draw another vertical line. The center of gravity of the figure is at the intersection of the two vertical lines. Other support points can be used to verify the position. The position can also be checked by supporting the figure from below with a finger or other small object. Verify that the center of gravity of a symmetric figure is at the geometric center.

3. Repeat Steps 1–2 uslng asymmetric figures, hollow figures, and human body shapes. Verify with a hollow figure or letter that the center of gravity can be in the space around the figure and not necessarily within the body of the figure. Have the students act out the human-body shapes and point out the positions of the center of gravity on the body.

4. Find the approximate location of the center of gravity for a flexible toy or doll. Loop a string around an arm or leg, suspend the toy, and mark or observe the location of the vertical relative to the toy. Suspend again from a different point and note the position of the center of gravity at the intersection of the two verticals. Bend the toy and repeat to verify that the center of gravity moves as the toy assumes different positions.

5. Find the approximate center of gravity of the wire-frame balance toy as in Step 4. Place the balance toy on its pedestal or on any level surface, and verify that its center of gravity is below the support point. Verify that it is in stable rotational equilibrium by displacing or pushing the toy and noting its return to equilibrium. Raise the center of gravity of the toy and convert this system to an unstable equilibrium by taping a dime near the top of the toy.

6. Use a clothespin as a skyhook to support leather belt. (See the Teaching Science with TOYS activity "Skyhook.") A skyhook is a folk toy used to support objects. Place the belt in the groove of the clothespin and balance it by placing a finger under the open end of the clothespin.

Part B: Human Balance Tricks

Have the students act out the following movements, first freestanding and then while touching a wall. In each case the wall prevents movement of the center of gravity over the support point, and the body cannot move to an equilibrium position.

1. Standing on toes—First, simply stand on your toes. Then face the wall with toes touching the wall and try to stand on your toes again. The wall prevents the movement of your center of gravity over your toes.

2. Leg lift—First, stand on your right foot while you lift your left leg. Then place your right side and the side of your right foot against the wall and try to lift your left leg. The wall prevents the movement of your center of gravity over the right foot.

3. Toe touching—First, touch your toes. Then place your back against the wall with your heels touching the wall and try to bend over and touch your toes. The wall prevents the movement of your center of gravity over your feet.

Variations and Extensions

Balance toy

1. Have students make balance toys using potatoes, wire coat hangers, corks, rubber stoppers, Styrofoam™ balls, and other items. (See photo.) Remind students that the center of gravity must be below the support point or the toy will fall over.

2. Illustrate that an object rotates about its center of gravity determined as in Part A of the "Procedure" by drawing circles centered at the center of gravity and circles centered elsewhere. Observe the circles when the figures are spun as they are thrown through the air.

3. Illustrate the shift of the center of gravity away from the center of a symmetric object by making a symmetric object into an asymmetric object. Use an object such as a plastic videotape case or a wooden or cardboard schoolbox and tape a small heavy item, such as a slotted weight, stone, or battery inside near a corner. If desired, drill holes through the box or case near the four corners, and locate the center of gravity as in the "Procedure."

4. Determine the approximate center of gravity of the human body: Have students make a mental note of the vertical symmetry line that bisects a human body. Have a student hang from a bar or rope with one hand. Note the vertical line formed by hanging. The intersection of the two lines is the center of gravity. (See Figure 1.)

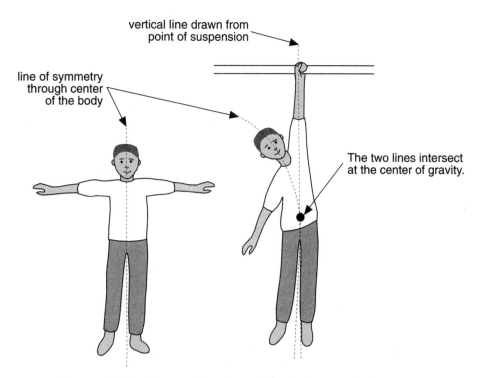

Figure 1: Find the center of gravity of a human body.

5. A commercially available balance toy known as a butterfly has small weights attached to the wings in order to make the toy more stable by lowering the center of gravity below the support point. This toy can be made from cardstock with coins taped or glued to the wings. (See Figure 2.)

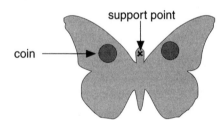

Figure 2: Use cardstock and coins to make a balancing butterfly.

6. Discuss these questions: How can a football player be made more stable? *By bending the knees (lowering the center of gravity) and spreading the feet further apart (widening the base).* How can an automobile be made more stable in a turn? *By lowering the center of gravity and lengthening the axle.*

Explanation

 The following explanation is intended for the teacher's information. Modify the explanation for students as required.

The force of gravity acts on each individual atom in an object. However, for most purposes, the force of gravity on the object, or weight, can be treated as a single force exerted at a point called the center of gravity of the object. A related concept, the center of mass, is almost identical to the center of gravity, and the difference is important only when parts of the object are at significantly different distances from the center of the Earth.

There are at least three ways to find the center of gravity: balancing the object from below, suspending the object from above, and calculating using the formula for the center of gravity. When the object is balanced or supported, it is in equilibrium and is acted on by two forces, the downward force of gravity and the upward balance or support force. These two forces must be equal in order to give zero net force as required for translational equilibrium, and they must act along the same line in order to give zero net torque as required for rotational equilibrium.

When the object is balanced from below (as on a finger), the center of gravity must lie directly above the support point (the point where the finger touches the object) or else the force of gravity would topple the object off the support. This can be shown by moving the finger to another point after balance has been achieved.

When the object is supported from above (as from a nail), the center of gravity must lie along a vertical line below the support point (along a string hanging from the nail) or else the force of gravity would rotate the object about this point. This can be shown by displacing the object to one side.

The formula for the location of the x component of the center of gravity x_{cg} of an object consisting of discrete parts is given by

$$X = \frac{w_1 x_1 + w_2 x_2 + w_2 x_3 \cdots}{W}$$

where the w's are the weights of the discrete parts, the x's are their x coordinates, and W is the total weight of the object. There is a similar formula for the location of the other components of the center of gravity.

For the special case of uniform, symmetric objects, the center of gravity lies at the geometric center. For other objects, the center of gravity is not at the center. The center of gravity does not even have to be within the material of the object. This can be shown by suspending an object in the shape of a letter such as O, C, or L from two positions and noting that the center of gravity must lie at the intersection of two vertical lines that cross outside the object. These aspects of the center of gravity can also be illustrated by calculating the location of the center of gravity of an object consisting of two 1-kg objects separated by 2 m:

$$x_{cg} = \frac{(1\ kg)(0\ m) + (1\ kg)(2\ m)}{2\ kg} = 1\ m$$

which is both at the center of the object and outside the object. (See Figure 3.)

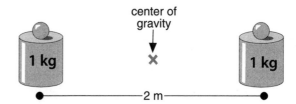

Figure 3: The center of gravity can be outside the object.

If an object is to be balanced, its center of gravity must be aligned vertically with the point of support, either above or below the point. If the center of gravity is above the support point, the equilibrium is classified as unstable since a small displacement will result in rotation of the object further away from equilibrium, caused by the force of gravity. If the center of gravity is below the support point, the equilibrium is classified as stable, since a small displacement will result in rotation by the force of gravity back towards equilibrium. On balance toys, small masses are added at low positions on the figures in order to lower the center of gravity below the support point and make the toys stable.

If, instead of being balanced on a point, an object is supported by a base, a vertical line through an edge becomes the dividing line between stable and unstable equilibrium, as shown in Figure 4. If the object is rotated so that its center of gravity does not go past this line, the object will return to its equilibrium position and become stable. If the object is rotated so that its center of gravity goes past this line, the object will rotate further from its equilibrium position and become unstable. The larger the angle through which an object must rotate to reach this line, the greater its stability. Two things can be done to increase this angle and therefore increase stability: lower the center of gravity and widen the base. A measure of stability is the ratio of the length of the base in the direction of the applied force and the height of the center of gravity. Stability can therefore be increased by widening the base and/or lowering the center of gravity.

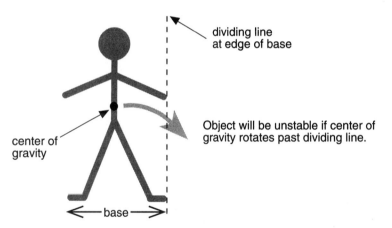

Figure 4: A vertical line through an edge of an object supported by a base becomes the dividing line between stable and unstable equilibrium.

References

Buckwash, V. "Ways to Demonstrate Center of Gravity," *Physics Teacher*. 1976, 14, 39.
Schnacke, D. *American Folk Toys;* Putnam: New York, 1973; p 64.

Contributor

Gary Lovely, Edgewood Middle School, Seven Mile, OH; Teaching Science with TOYS peer mentor.

WHIRLING STOPPER

A few ordinary materials become an intriguing toy that introduces conservation of angular momentum.

Whirling Stopper

GRADE LEVELS

Science activity appropriate for grades 7–9
Cross-Curricular Integration intended for grades 7–9

KEY SCIENCE TOPICS

- angular momentum
- centripetal force
- momentum

STUDENT BACKGROUND

Students should be familiar with the topics of force, speed, and inertia.

KEY PROCESS SKILLS

• observing	Students observe the motion produced by a centripetal force using a homemade whirling toy.
• measuring	Students time revolutions of the stopper in order to calculate speed for calculation of centripetal force.

TIME REQUIRED

Setup	5	minutes
Performance	15	minutes
Cleanup	5	minutes

Materials

For the "Procedure"
Per class

- 1-ounce fishing sinkers to be used as weights
- 15-cm-length of ¾-inch PVC pipe that the sinker will pass through
- (optional) solvent such as methyl ethyl ketone

PVC pipe and methyl ethyl ketone are available from hardware stores.

- approximately 1.5-m length of strong thin nylon cord or fishing line
- 1 2-hole #3 rubber stopper
- 1 sheet of fine sandpaper
- stopwatch

For "Variations and Extensions"

❶ Per class
- Whirling Stopper apparatus (made in "Getting Ready")
- more sinkers

❷ Per class
- Whirling Stopper apparatus (made in "Getting Ready")
- (optional) a rotating stool or Sit 'N Spin®
- (optional) hand weights

❸ Per class
- pennies and balloons, BBs or ball bearings and a flask or funnel, or Vortex® and coin

 Vortex® is available from Divnick International, Inc.; Spring Valley, OH 45370.

Safety and Disposal

Methyl ethyl ketone is a flammable organic solvent. Use in a well-ventilated area. Use only according to manufacturers' specifications and follow the safety precautions listed on the container.

Ample space must be allowed for the spinning of the apparatus. The condition of the cord should be checked before each use to protect against the stopper becoming a dangerous projectile. Careful smoothing of the pipe as discussed in "Getting Ready" will help keep the pipe from cutting the cord as it moves.

No special disposal procedures are required.

Getting Ready

Construct a Whirling Stopper:

1. Cut a 15-cm piece of ¾-inch PVC pipe. Sand one end to a very smooth surface. (If desired, further smooth the sanded end by dipping in an appropriate solvent such as methyl ethyl ketone.)

2. Tie one end of the cord to the stopper and the other end to the sinker.

3. Pass the sinker through the pipe so that the stopper end of the cord is at the sanded end of the pipe.

Procedure

1. Hold the pipe in one hand. Allow approximately 50 cm of the stopper end of the cord to extend out of the pipe. The sinker end of the cord will hang from the other end.

2. As you begin to discuss the activity with the class, rotate the stopper end of the tube slowly until the stopper whirls in a horizontal circle over your head. Continue whirling the stopper, using this time to review previous material such as rotation, orbits, period, force, and speed that will support this lesson.

3. Gradually increase the rotational speed until the sinker is about halfway between its lowest position and the lower edge of the pipe. Now, try to maintain a constant velocity to allow the system to achieve equilibrium.

 The weight of the sinker provides the required centripetal force to keep the stopper moving in a circle.

4. Calculate the speed of the stopper:

$$v = \frac{2\pi r}{T}$$

where r is the radius of the circle from the edge of the pipe to the center of the stopper and T is the period or time for one revolution. (Determine T by timing three to five revolutions of the stopper and finding the average time.)

5. Calculate the centripetal force:

$$F = \frac{mv^2}{r}$$

where m is the mass of the stopper. Compare with the weight of the sinker:

$$W = Mg$$

where M is the mass of the sinker. Note that if the stopper is whirled faster, at equilibrium the stopper will move to a larger radius to maintain:

$$F = W$$

6. While whirling the stopper at a constant speed, pull the sinker downward and allow the students to observe that the speed increases as the radius of the stopper's circle decreases, in accordance with the Law of Conservation of Angular Momentum.

Variations and Extensions

1. Further studies can be done by adding more sinkers to increase the weight and centripetal force and repeating the "Procedure."

2. To verify the Law of Conservation of Angular Momentum, whirl the stopper to the previous radius of known speed. Then pull the sinker end of the cord down to some predetermined radius, hold it there, and quickly time as many revolutions as possible to obtain the new speed before friction slows the stopper. The conservation law requires that the angular momentum of the stopper, given by

$$mvr$$

be the same at the old and new positions. Compare the Whirling Stopper to an ice skater whose speed increases as a result of pulling in her arms to a new radius. A rotating stool or Sit 'N Spin® with hand weights can be used to demonstrate this effect.

3. The centripetal force on an orbiting object and the Law of Conservation of Angular Momentum can be demonstrated by whirling a penny on its edge in an inflated balloon, a BB or ball bearing in a glass flask or funnel, or a coin in a Vortex®. In each case, as the orbiting object moves down the cone or slope, the

radius decreases, and the speed increases. If energy is fed into the system by rotating the balloon, funnel, or base of the Vortex® at the resonant frequency of the orbiting object, the object will maintain or increase its radius.

Explanation

 The following explanation is intended for the teacher's information. Modify the explanation for students as required.

For any object to move in a circle, there must be a force on the object directed along a radius toward the center of rotation. If this force, called the centripetal (center-seeking) force, is not present, the object simply will not move in a circle. The centripetal force can be supplied by any contact force or action-at-a-distance force, such as gravity or the electrical force. Examples are the force of the cord on a yo-yo as it goes through a loop, the force on the driver of a car from seat belts or the inside of the door as the car turns a corner, the gravity force of the Earth on the moon or the sun on the Earth as these objects revolve in orbits, and the electrical force of the proton on the electron in the Bohr model of hydrogen. If the centripetal force is removed while an object is moving in a circle, such as by cutting the cord of the Whirling Stopper, the object will continue motion in a straight line in the direction of its tangential velocity.

The magnitude of the centripetal force F required for an object of mass m to move at speed v in a circle of radius r is given by

$$F = \frac{mv^2}{r}$$

According to Newton's Second Law,

$$F = ma$$

Thus, the magnitude of the centripetal acceleration is given by

$$\frac{v^2}{r}$$

The direction of the centripetal acceleration is the same as the direction of the centripetal force, which is towards the center of rotation.

For the Whirling Stopper, the centripetal force is provided by the tension in the cord that is passed down through the PVC pipe and attached to the sinker. The tension in the cord results from the weight of the sinker, and the lip of the PVC pipe acts as a pulley to change the direction of the tension from vertical to horizontal. The centripetal force F is equal to the weight:

$$W = Mg$$

of the sinker, where M is the mass of the sinker. The expression

$$F = W$$

can be verified experimentally by making the measurements necessary to calculate the centripetal force and measuring the weight.

For a horizontal orbit, the angular momentum of an object is given by

$$\text{Angular Momentum} = mvr$$

where mv is the linear momentum and r is the radius of the orbit. If there are no forces turning the object about an axis perpendicular to the plane of the orbit, the angular momentum about that axis is constant. This is the Law of Conservation of Angular Momentum. The Whirling Stopper should obey this law because the tension in the cord does not turn the stopper about the axis along the PVC pipe but instead is directed toward the axis. The Law of Conservation of Angular Momentum in the form

$$m_1 v_1 r_1 = m_2 v_2 r_2$$

can be verified experimentally by measuring the speed and radius at two positions, 1 and 2.

Cross-Curricular Integration

Earth science:
- Have the students study planetary orbits.

Home, safety, and career:
- Have the students investigate what forces are involved in a car going around a curve in the road.

Physical education:
- Discuss why ice skaters, divers, and gymnasts spin faster when they pull their arms inward.

References

Physical Science Study Committee. *College Physics Laboratory Guide;* Raytheon Education: 1968.

Trinklein, F.E. *Modern Physics;* Holt, Rinehart and Winston: Austin, TX, 1992.

Contributor

Gary Lovely, Edgewood Middle School, Seven Mile, OH; Teaching Science with TOYS peer mentor.

FLOATING CANS

Students are introduced to Archimedes' principle by observing how mass, density, and volume affect whether cans float or sink in water.

Cans in aquarium

GRADE LEVELS

Science activity appropriate for grades 7–9
Cross-Curricular Integration intended for grades 7–9

KEY SCIENCE TOPICS

- Archimedes' principle
- buoyant force
- density
- mass
- sinking and floating
- volume

STUDENT BACKGROUND

Students should have a thorough knowledge of density and mass and be able to measure or calculate the volume of regular and irregular objects.

KEY PROCESS SKILLS

- predicting — Students predict whether soft-drink cans will sink or float.

- investigating — Students investigate mass, density, and volume concepts using various floating and sinking soft-drink cans.

TIME REQUIRED

Setup	10 minutes
Performance	25 minutes
Cleanup	5 minutes

Materials

For the "Procedure"
Per group of 4 students
- 2½-gallon bucket
- spring scale
- equal-arm balance and weights

- various objects with densities greater than and less than that of water, such as cork, metal, candle, wood, rubber stopper, and a very dense wood (like ironwood or ebony, if available)
- 1-L graduated cylinder
- ruler

Per class
- 10-gallon aquarium
- assorted soft drinks (some artificially and some sugar-sweetened) in 12-ounce aluminum cans
- 8 gallons water
- 4 cups salt

 Kosher salt is preferable to table salt, as additives in table salt cloud the solution.

For the "Variations"

❶ Per class
- assorted drinks in 12-ounce cans, including at least 1 of each type listed below:
  - artificially sweetened
  - sugar-sweetened
  - carbonated
  - uncarbonated
  - bi-metal can
  - aluminum can

 Aluminum cans can be distinguished from other types by the smooth and shiny appearance of the bottoms and by the fact that they are not attracted to magnets.

Safety and Disposal

No special safety or disposal procedures are required.

Procedure

Part A: Exploring Buoyancy

1. Ask the students to predict whether the different objects will float or sink. Have them test objects made of cork, metal, and wood. The students probably predicted these correctly. Then have them test a candle, a rubber stopper, and a very dense wood. They are less likely to have predicted these correctly.

2. Direct the students toward the conclusion that the objects that sank have densities greater than water and the ones that floated have densities less than water.

3. Reinforce the fact that water has a density of 1 g/cm³. That is, each cubic cm of water has a mass of 1 g.

4. Have students use a spring scale to measure the weight of one of the objects from Step 1 out of water and then submerged under water. Ask the students why there is a difference.

5. Discuss the buoyant force.

6. Measure the volume of the object used in Step 4 by placing the object in the graduated cylinder and observing the difference in the height of the water. Discuss displaced water. On the board write the volume of the object and multiply it by the density of water to get the mass, then by the acceleration of gravity to get the weight. Show that the object in water is lighter than it is out of water by the weight of the displaced water. (This difference in weight is the buoyant force acting on the object.)

7. Measure the mass of the objects used in Step 1 on the balance and measure their volume by water displacement or by measuring with a ruler. Calculate their densities (mass divided by volume) and verify that sinking/floating occurs for objects with densities greater/less than the density of water (1 g/cm³).

Part B: Floating Cans

1. Place the aquarium where all students can see it. Fill the aquarium with water.

2. Review the definition of density. Work several problems on the board as part of the review. Take care not to mention materials with densities less than that of water.

3. Hold one can of artificially sweetened and one can of sugar-sweetened soda in the aquarium. Ask the students whether each will sink or float.

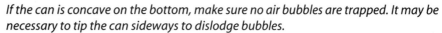

If the can is concave on the bottom, make sure no air bubbles are trapped. It may be necessary to tip the can sideways to dislodge bubbles.

4. Release the cans.

5. Ask the students why the sugar-sweetened soda sank.

6. Elicit the response that the can that sank was more dense. Seek an explanation for the difference in density.

7. After the students understand that the sugar must have increased the density, ask why one floated and one sank in water.

8. Direct the students toward the conclusion that the one that sank has a density greater than water and the one that floated had a density less than water. Reinforce the fact that water has a density of 1 g/c³.

9. Have students measure the mass and volume of the floating and nonfloating cans and calculate their densities. Have them compare the densities of the cans to the density of water.

10. Test cans of various beverages to determine whether they float or sink in water.

11. Remove the cans from the tank.

12. Add 2 cups of kosher salt to the water. Stir until dissolved.

13. Test the cans again to see if any of the cans that previously sank will now float.

14. Repeat Steps 12 and 13.

Variations

1. To take a learning cycle approach to this lesson, do a slightly different version of Part B before doing Part A and before having any discussion of density or buoyancy. Before beginning the demonstration, consider the order in which you will add the cans. Put in a few cans that will lead the students to a particular conclusion and then add one that will give a different result. For example, put in several cans of diet soda that float and then add one in a bimetal can that sinks. (Be sure to test in advance.)

   Do not reveal the variables; let the students figure out what they are. As students recognize the different variables involved (can material, sugar/diet, caffeine/decaffeinated, carbonated/uncarbonated), list them on the board or on an overhead. As the demonstration proceeds, students can then check categories, keeping track of trends revealed by the data. This approach allows students to develop and modify hypotheses about the floating/sinking of cans as the activity progresses.

2. Demonstrate the following: If a can that does not float is dropped on the floor from about 2 feet, it will expand slightly. This small increase in volume can decrease the density enough so that the can will float.

Explanation

The following explanation is intended for the teacher's information. Modify the explanation for students as required.

When an object is submerged, water exerts an upward force that is opposite in direction to the force of gravity. This upward force is called the buoyant force. An object floats if the buoyant force acting on the submerged object is equal to the weight of the object. The size of the buoyant force depends on the amount of liquid displaced by the submerged object. The relationship between buoyant force and displaced liquid was described about 2,300 years ago by the Greek philosopher Archimedes and is called Archimedes' principle. It states that an object that is floating or submerged in a liquid is buoyed upward by a force equal to the weight of the displaced liquid. (See Figure 1.)

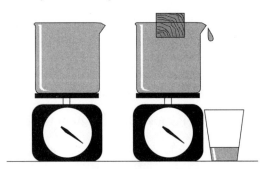

Figure 1: A floating object displaces a weight equal to its own weight.

As an object is lowered into the liquid, the buoyant force increases; it reaches a maximum when the object is completely submerged. This maximum force is equal to the weight of the volume of liquid displaced. The volume of liquid displaced is equal to the total volume of the object.

However, we do not have to calculate maximum buoyant force to determine if an object will float; we can simply compare the density of the object to the density of the liquid. As we have discussed, the net force acting on an object placed in a liquid depends on the relative weight of equal volumes of the object and the liquid. With equal volumes, greater density means greater weight. Thus, we can simply compare the density of the object to the density of the liquid. Therefore, an object floats whenever the density of the object is less than the liquid's density.

In Part B, the volume displaced is very nearly equal for all cans, resulting in equal buoyant forces. Therefore, the determining factor in sinking or floating is the mass of the full cans. The mass depends on three variables: the amount of the soda in the can, the density of the soda, and the mass of the metal from which the can is made. Soda sweetened with sugar has a greater density than soda with artificial sweetener because a greater amount of sugar is required to sweeten the soda than the amount of artificial sweetener needed for the same amount of soda. Steel or bimetallic cans have more mass than aluminum cans because steel is more dense than aluminum. The densities of soda cans are so nearly equal to the density of water that variations in the amount of soda in the can might cause unexpected behaviors.

Cross-Curricular Integration

Social studies:
- Have students research the size of historic sailing ships and the types of cargo they carried. Direct the students to consider how the density of the cargo would affect the amount that could be safely carried.

References

Kirkpatrick, L.D.; Wheeler, G.F. *Physics: A World View;* Saunders College: Ft. Worth, TX, 1992; pp 166, 175.

Tropp, H.E.; Friedi, A.E. *Modern Physical Science;* Holt, Rinehart and Winston: New York, 1991; p 345.

Contributors

Gary Lovely, Edgewood Middle School, Seven Mile, OH; Teaching Science with TOYS peer mentor.
Tom Runyan, Garfield Alternative School, Middletown, OH; Teaching Science with TOYS peer mentor.

FALLING FILTERS

Students discover that drag affects falling objects.

Falling filters

GRADE LEVELS

Science activity appropriate for grades 7–12
Cross-Curricular Integration intended for grades 7–9

KEY SCIENCE TOPICS

- acceleration
- drag force
- terminal velocity
- velocity

STUDENT BACKGROUND

Students should thoroughly understand acceleration due to gravity.

KEY PROCESS SKILLS

• observing	Students observe the effect of air resistance on falling objects.
• collecting data	Students collect and compare data to discover how drag affects falling objects.

TIME REQUIRED

Setup	5	minutes
Performance	15	minutes
Cleanup	5	minutes

Materials

For "Introducing the Activity"
- small objects such as the following:
    - coins
    - erasers

For the "Procedure"
Per group of 4–5 students
- 1 meterstick
- 5 clothespins
- 5 coffee filters

For the "Extensions"

❷ Per class
- 1 Styrofoam™ ball, 1 inch in diameter
- 1 Styrofoam™ ball, 3 inches in diameter

Safety and Disposal

If a student must stand on a table to reach the drop height, have another student assist. No special disposal procedures are required.

Getting Ready

Sets of measured height marks could be placed around the room on the walls as drop points. Display Table 1 from the "Explanation" on the board or in a handout as part of a data sheet.

Introducing the Activity

Have the students drop several small objects, such as coins or erasers, for which air resistance is negligible, and lead them to the conclusion that in the absence of air resistance, all objects fall at the same rate. Then give them a coffee filter and let them discover that this is an object for which air resistance is not negligible. Lead them to the concept of drag force as a name for air resistance and the concept of terminal velocity when drag force equals weight.

Lead the students to the conclusion that there is zero drag force when an object is at rest and that the drag force increases as the velocity increases and tell them that this activity will investigate the dependence of the drag force on velocity. Discuss with them the idea that the drag force can be studied for objects such as coffee filters that have a shape and density that enable them to reach terminal velocity very quickly after they are released.

Next discuss skydivers and relate their behavior to that of the coffee filters. Lead them to contrast the behavior of a skydiver before and after the parachute opens. Before the parachute opens, the velocity of the skydiver increases to very large values, in the 100–200 miles-per-hour range, before air resistance equals weight and terminal velocity is reached. After the parachute opens, the large drag force of the parachute results in a relatively low terminal velocity. If the skydiver was dropped with his or her parachute open, the velocity would again increase until the appropriate terminal velocity was reached.

Discuss the classic demonstration of a penny and a feather falling in a glass tube that is first filled with air and then evacuated. This demonstration was repeated by the astronauts on the moon with a feather and a hammer. Discuss why the feather and penny or feather and hammer fall at the same rate in the absence of air.

Procedure

Divide the students into groups of four or five and assign jobs.

Each group will need a Recorder, two Droppers, and one or two Watchers to perform the following jobs:
- *Recorder—to record the group's results;*
- *Dropper—to drop the objects; and*
- *Watcher—to watch the objects drop and determine if they hit the ground simultaneously.*

Part A: Dropping Objects from Equal Heights

1. Have Droppers practice dropping objects from both hands at the same time.

Throughout the activity, the Dropper will be releasing one or more objects simultaneously from each hand.

2. Have the Watchers practice observing the objects as they hit the ground.

3. Instruct Droppers to hold one clothespin in one hand and four clothespins (clipped together) in the other at a height of 1 m and release simultaneously.

4. Have Watchers determine whether or not the clothespins hit the ground simultaneously.

5. Tell Recorders to record the groups' results.

6. Have groups repeat Steps 3–5, this time holding clothespins at a height of 2 m.

7. Instruct groups to repeat Steps 3–5, this time replacing the clothespins with one coffee filter (open side up) in one hand and four nested coffee filters in the other.

8. Have groups repeat Step 7, this time holding coffee filters at a height of 2 m.

9. Discuss students' observations. Did the clothespins hit the ground at nearly the same time? *Yes.* How about the coffee filters? *No.* Why were the coffee filters different? *Because of their shape and density, drag had a more noticeable effect on the coffee filters than on the clothespins. Because the coffee filters experienced significant drag, the difference in weight between the two sets of filters caused an observable difference in drop rates.*

Part B: Dropping Filters from Different Heights

1. Have each Dropper hold one coffee filter at a height of 1 m and two coffee filters nested together in the other hand at a height of 2 m.

2. Have the Droppers release the filters from both hands simultaneously.

3. Instruct the Watchers to determine whether or not the filters hit the ground simultaneously (or approximately simultaneously).

The reason for doing this is to determine equal drop times, as noted in the "Explanation."

4. Tell Recorders to record the groups' results on the Data Sheet (provided). Have the Recorders write "yes" for the trial if the filters landed at the same time and "no" if the filters did not land at the same time.

5. Have groups complete two more trials by repeating Steps 1–4 two more times.

6. Have groups repeat the experiment as in Steps 1–5 but vary the height of the 2-filter unit as follows: 1.9 m, 1.8 m, 1.7 m, 1.6 m, 1.5 m, 1.4 m, 1.3 m, 1.2 m, 1.1 m, and 1 m.

7. Introduce the meaning of the exponent "α" relating to drag and velocity. (See "Explanation" for a complete discussion.)

8. Instruct the groups to look on Table 2 on the Data Sheet for the height closest to the one for which they observed the filters hitting the ground simultaneously. Have groups record the corresponding value of α.

9. Compare data of all groups on the board and discuss the significance of α and the relationship of drag and velocity.

Extensions

1. Repeat Part B with a different number of filters. For example, with four filters $(W_2/W_1) = 4$ in equation 5 in the "Explanation," and this yields $H_2 = 2$ for $\alpha = 2$.

2. Drop Styrofoam™ balls 1 inch and 3 inches in diameter from an elevated position, such as by standing on a table. Before dropping, ask students to predict which ball will hit the floor first, the 1-inch ball because it has less cross-sectional area and less drag or the 3-inch ball because it has more mass. *The 3-inch ball wins because the terminal velocity of balls having the same density is proportional to the cross-sectional area (radius squared).*

Explanation

The following explanation is intended for the teacher's information. Modify the explanation for students as required.

Objects falling through a fluid, liquid or gas, experience an opposing frictional force known as drag. As the object begins to fall, this force is small. As the object speeds up, the drag also increases. The maximum velocity of the object is reached when the weight of the object is equal to the drag. The maximum velocity of a falling object is known as its terminal velocity. Objects have different terminal velocities because they have different weights and therefore attain different velocities before drag increases enough to equal weight. Because objects have different terminal velocities, they will also have different fall times. This difference in fall times is not noticeable for objects like clothespins because the drag is so small compared to the weight that they never reach terminal velocity. In fact, their acceleration is very nearly equal to the acceleration due to gravity (9.8 m/s^2) for the entire fall.

In many situations, the drag force is proportional to the velocity raised to some power. Therefore, the drag force, D, can be expressed as a constant, c, times the velocity, v, of the object raised to some exponent α, which usually varies between 1 and 2 depending on the velocity range, size, shape, orientation, and surface condition of the object:

(1) $$D = cv^\alpha$$

where $\alpha = 1$ for a linear dependence, $\alpha = 2$ for a squared dependence, etc. When drag equals weight, W, the velocity equals the terminal velocity, V:

(2) $$D = W = cV^\alpha$$

Assuming that an object dropped from height H reaches terminal velocity very quickly after it is dropped:

(3)
$$H = VT$$

where T is the time of fall. The ratio of two drop heights can be expressed as

(4)
$$\frac{H_2}{H_1} = \frac{V_2 T_2}{V_1 T_1}$$

If objects of different weights are dropped from different heights and have the same drop times because of different drags, from equations (2) and (4):

(5)
$$\frac{H_2}{H_1} = \frac{V_2}{V_1} = \left[\frac{W_2}{W_1}\right]^{\frac{1}{\alpha}}$$

An approximate value of the exponent, α, relating drag and velocity can be found by measuring $H_1 = 1$ m and selecting $W_2 = 2\,W_1$ by letting W_1 be one filter and W_2 be two identical filters:

(6)
$$H_2 = (2)^{\frac{1}{\alpha}}$$

Let α be an integer, so that $H_2 = 2$ m for $\alpha = 1$; $H_2 = 1.41$ m for $\alpha = 2$; $H_2 = 1.26$ m for $\alpha = 3$; etc. A measured value of H_2 near one of these values yields a value of α to the nearest integer.

Cross-Curricular Integration

Social studies:
- Discuss or have students write reports on the development and impact of the parachute. List humanitarian, military, and recreational uses of the parachute.

Reference

Kirkpatrick, L.D.; Wheeler, G.F. *Physics: A World View;* Harcourt, Brace, and Jovanovich: Orlando, FL, 1992.

Contributor

Gary Lovely, Edgewood Middle School, Seven Mile, OH; Teaching Science with TOYS peer mentor.

Handout Master

A master for the following handout is provided:
- Data Sheet

Copy as needed for classroom use.

FALLING FILTERS
Data Sheet

1. Do the falling filters hit the ground simultaneously? Write the answers for each trial in the table below.

Table 1				
		Do they hit the ground at the same time? (Yes or No)		
Height of single filter	Height of 2-filter unit	Trial 1	Trial 2	Trial 3
1 m	2 m			
1 m	1.9 m			
1 m	1.8 m			
1 m	1.7 m			
1 m	1.6 m			
1 m	1.5 m			
1 m	1.4 m			
1 m	1.3 m			
1 m	1.2 m			
1 m	1.1 m			
1 m	1 m			

2. Look on Table 2 for the height closest to the one for which you observed the filters hitting the ground simultaneously, and find the corresponding value of α.

Table 2	
Height (m)	possible values of α
2	1
1.41	2
1.26	3
1.19	4

STICK AROUND

Students explore why a boomerang "sticks around."

GRADE LEVELS

Science activity appropriate for grades 7–9
Cross-Curricular Integration intended for grades 7–9

KEY SCIENCE TOPICS

- gyroscope
- lift force
- relative speed
- rotation

Boomerangs: a) three-arm, b) crossed-arm, c) NASA

STUDENT BACKGROUND

Students should be familiar with the lift force resulting from the flow of air over a wing or blade and with the ability of a force to cause the rotation of an object.

KEY PROCESS SKILL

- observing

Students observe the behavior of various shapes of boomerangs.

TIME REQUIRED

Setup	10	minutes
Performance	50	minutes
Cleanup	5	minutes

Materials

For the "Procedure"

Per student

- utility knife with retractable blade
- 30-cm square of thin corrugated cardboard

Pizza boxes work nicely. Many pizza restaurants will sell new boxes. The weight of Pizza Hut boxes is ideal for making boomerangs. The side strips of the boxes work well for the crossed-arm boomerang. (See Extension 2.)

For "Variations and Extensions"

❶ All materials listed for the "Procedure" except
- substitute a piece of cardboard with a different thickness

❷ Per student
- stapler
- 2 strips of thin cardboard

 The strips of cardboard can be any size, but 3 cm x 30 cm works nicely.

❸ Per class
- bicycle wheel with an axle
- 1 2-foot length of rope

Safety and Disposal

Instruct students to always direct a sharp blade away from their bodies and others. No special disposal procedures are required.

Procedure

Have students do the following:

1. Cut out the Three-Arm Boomerang Pattern (provided). Trace it onto the cardboard, keeping one boomerang blade parallel to the corrugations. Use the utility knife to cut out the boomerang.

2. Curve each of the three blades up slightly, keeping the center flat.

3. Hold the boomerang loosely between the thumb and forefinger and tip it slightly to the right (if right-handed). Throw the boomerang overhand like a pitcher throwing a baseball by snapping the wrist and releasing with the arm extended at or slightly above eye level.

4. If necessary to improve performance, bend about 1 cm of the leading edge of each blade upwards about 30° in order to produce lift on the blade. (See pattern.)

If the boomerang is thrown into the wind, throw at 45° to the right of the wind direction, when facing into the wind (if right-handed).

5. Observe the results of your throw and adjust the bends as needed.

The more bend, the tighter the circle the boomerang makes.

6. Discuss the behavior of the boomerang and variations when changes are made.

Variations and Extensions

1. Once students have developed a suitable throwing technique that yields consistent results, they could modify the boomerang made in the "Procedure." For example, students could vary the size of the boomerang or the thickness of the cardboard.

2. Students could also try alternate boomerang shapes, such as the crossed-arm or NASA boomerangs. The NASA boomerang shown in the photograph on the first page of this activity is a replica of the boomerang used in the NASA Toys in Space 2 program flown on board Space Shuttle Flight STS-54 in 1993. It has a diameter of 18 cm and is made from cardstock. Use the pattern at the end of this activity to make a NASA boomerang. A crossed arm boomerang is made as follows: staple two pieces of thin cardboard together perpendicular to each other. The strips of cardboard can be any size, but 3 cm x 30 cm works nicely. The blade tips can be bent upward to produce lift.

3. As mentioned in the "Explanation," the boomerang behaves very much like a gyroscope. Make a gyroscope to illustrate this behavior from a bicycle wheel with an axle to which a short piece of rope is tied. Have a student stand on each side of the wheel and hold the axle horizontally. Have one student start the wheel spinning. The student on the rope side of the wheel should then release the axle and hold the rope. When the other student also releases the axle, the wheel revolves or precesses in a circle keeping its axle horizontal, just like the boomerang.

Explanation

 The following explanation is intended for the teacher's information. Modify the explanation for students as required.

When the boomerang is thrown (right-handed), it is spinning about a horizontal axis with the upper blade moving forward and the lower blade moving backward. This combination of forward translational motion plus forward rotational motion gives the upper blade a larger speed relative to the air than the lower blade. This larger speed gives the upper blade more lift than the lower blade. The lift in this case is perpendicular to the blade and directed to the thrower's left.

These unbalanced lift forces attempt to rotate the boomerang from the vertical plane in which it is spinning to a horizontal plane. However, since the boomerang is spinning, it does not move toward a horizontal position but instead rotates its left face towards the thrower (right-handed) keeping its spin axis horizontal. Since the unbalanced lift forces act continuously, the boomerang continues to rotate in this manner and follows a circular path back to the thrower. (See Figure 1.)

The spinning boomerang behaves very much like a spinning gyroscope, and the rotation of the boomerang is known as gyroscopic precession. In addition, the stability of the spinning boomerang against unwanted motions caused by small forces is known as gyroscopic stability.

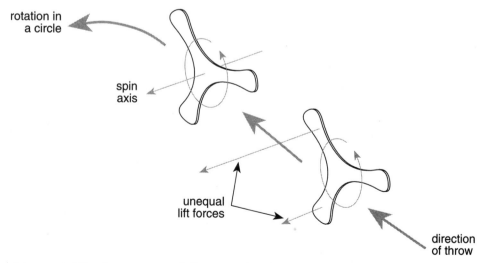

Figure 1: The boomerang follows a circular path back to the thrower.

Cross-Curricular Integration

Language arts:
- Have students research and write reports about the origin and past and present uses of the boomerang.

References

Darnell, E.; Ruhe, B. *Boomerang—How to Throw, Catch and Make It;* Workman: New York, 1985; pp 23–24, 47–48.

Trinklein, F.E. *Modern Physics;* Holt, Rinehart and Winston: Austin, TX, 1992.

Contributors

James Helwig, Edgewood Middle School, Hamilton, OH; Teaching Science with TOYS, 1990–91.

Gary Lovely, Edgewood Middle School, Hamilton, OH; Teaching Science with TOYS peer mentor.

Handout Masters

Masters for the following handouts are provided:
- Three-Arm Boomerang Pattern
- NASA Boomerang Pattern

Copy as needed for classroom use.

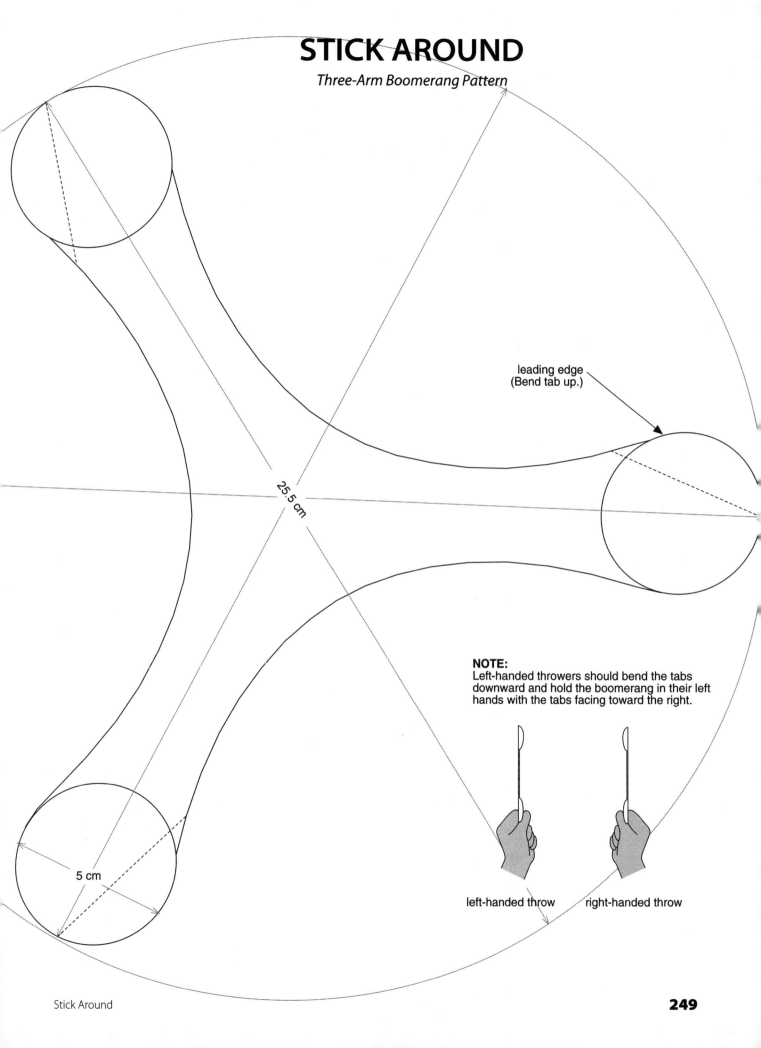

STICK AROUND
Three-Arm Boomerang Pattern

leading edge
(Bend tab up.)

25.5 cm

5 cm

NOTE:
Left-handed throwers should bend the tabs
downward and hold the boomerang in their left
hands with the tabs facing toward the right.

left-handed throw

right-handed throw

STICK AROUND

NASA Boomerang Pattern

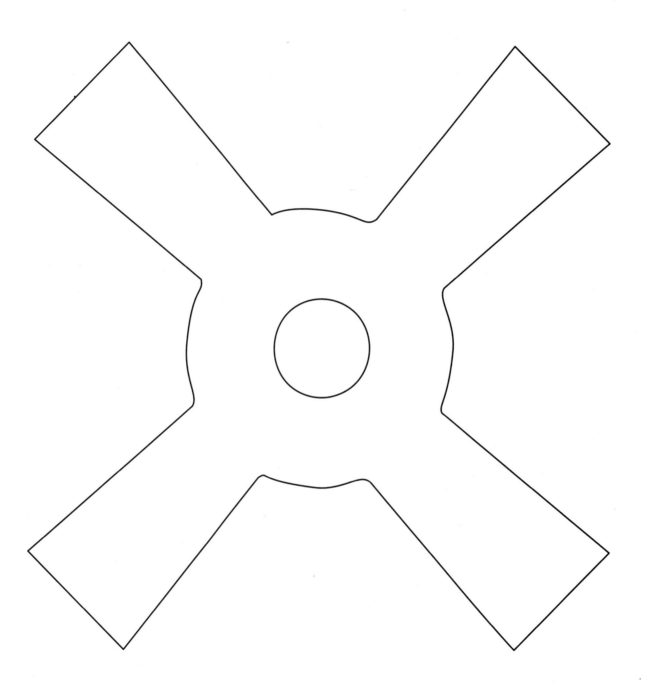

DELTA DART

Students' enthusiasm levels soar as they explore the principles of flight while building and experimenting with the Delta Dart.

Delta Dart

GRADE LEVELS

Science activity appropriate for grades 7–9
Cross-Curricular Integration intended for grades 7–9

KEY SCIENCE TOPICS

- force of gravity
- lift
- Newton's Third Law
- thrust

STUDENT BACKGROUND

Students should be familiar with the basic concepts of gravity, center of gravity, and Newton's Third Law.

KEY PROCESS SKILLS

• predicting	Students predict the effect of design modification on the flight of an airplane.
• controlling variables	Students make adjustments to their airplanes in order to achieve optimal flying characteristics.
• investigating	Students use a model airplane to investigate the principles of flight.

TIME REQUIRED

Setup	30	minutes
Performance	90	minutes
Cleanup	5	minutes

Materials

For the "Procedure"
Per student
- 1 Delta Dart kit

Delta Dart kits are available from Midwest Products Co., Inc.; 400 S. Indiana Street; P.O. Box 564; Hobart, IN 46342; (219) 942-1134.

- 1 approximately 60-cm x 50-cm piece of corrugated cardboard to serve as a pinning surface
- wood glue
- small disposable cup
- 8 straight pins
- single-edge razor blade

Per class
- hot glue gun and glue or epoxy glue

Epoxy works better, but is slower and messier.

- 1 roll of 1-inch-wide self-adhesive brown mailing tape
- Superglue

The Superglue is for emergency repairs to the airplanes in the field.

For "Variations and Extensions"
All materials listed for the "Procedure" plus the following:

❶ Per class
- several tape measures or metersticks

❸ Per class
- several coins to use as weights

❹ Per class
- 2-m string
- pulley
- standard weight set that can be attached to a string

Safety and Disposal

Instruct students to use extreme care with razor blades. You should be the only one to use the hot glue gun. No special disposal procedures are required.

Getting Ready

Build one airplane in advance to become familiar with the kit. If the airplanes are to be built over a period of time, arrange storage since the unfinished projects are rather bulky and fragile.

Set up a station for each student as follows: From the Delta Dart kit, give each student nine small balsa sticks, one fuselage stick, one tissue plan sheet, and one set of instructions. Also give each student a small amount of glue in a small plastic cup and eight pins.

Procedure

1. Have the students construct Delta Darts using the instruction sheet included in the kit.

   ➤ *To avoid running out of long balsa sticks, make sure students construct the horizontal stabilizers prior to construction of the wing. Tell students to conserve sticks where possible by using short pieces. Note that this order of construction may differ from some versions of the Delta Dart instruction sheet.*

2. When the wings are on and the airplane is pinned to the cardboard, have the students bring their projects to you to hot glue or epoxy. Allow hot glue to cool.

   ➤ *For firmer construction, allow the airplanes to dry overnight.*

3. Pass out rubber bands and show students where to place the pin to hold the end of the rubber band.

4. Distribute about 6 inches of brown mailing tape to each student and have the students place masking tape tabs on the control surfaces as shown in the plans.

5. Have students put their names or initials on their planes.

6. Take the students and planes outside. For the initial flight test, have the students wind the propeller about 75 turns.

   ➤ *Take an emergency repair kit into the field consisting of Superglue, a number of short balsa wood sticks, pins, and a few rubber bands and propellers.*

7. After the flight, instruct the students to adjust the control surfaces to be correct for level flight if required; for example, if the plane is climbing, bend the control on the horizontal stabilizer down a small amount and try again. Have the students continue adjustments until level flight is attained.

8. After a period of trial flights, have the students wind the propeller about 200 times. Students should line up facing the wind and launch the planes one at a time. Have each student stand by his or her plane at its landing point. Acknowledge the "winner" and "runners-up."

9. Have students prepare to launch planes as in Step 8, and conduct a competition for longest flight time. Acknowledge the winner and runners-up. Repeat the flight, this time racing for longest flight distance in any direction (chosen individually by each student).

10. (optional) For a nice finish, have all students launch their planes at the same time.

Variations and Extensions

1. Once students have achieved stable flight, try a series of numbers of turns of the propeller (such as 100 turns, 200 turns, etc.). Measure the distance flown by each. Plot a graph of turns versus distance.

2. As a class, choose a well-trimmed (smooth-flying) airplane and launch it from a given height with a given force (number of turns) and determine the glide slope in free flight (drop versus distance flown).

3. Do Extension 2, but vary the weight by placing coins at the airplane's center of gravity and determine the glide slope.

4. As a class, determine the static thrust produced by the rubber band and the propeller by tying a string to the back of the plane, positioning the string over a pulley, and attaching weights to the string until the plane doesn't move.

Explanation

 The following explanation is intended for the teacher's information. Modify the explanation for students as required.

The successful flight of an airplane depends on its being able to lift off the ground and to be controlled during flight. Although the Delta Dart airplane has a mass of only a few grams, the physical principles that enable it to fly are the same as those for a large jet flying across continents.

When an object such as an airplane moves through a fluid like air, the air exerts a force on the object. The component of this force that is at a vertical right angle to the direction of motion is called the lift force. During the level flight of an airplane, the lift force opposes the force of gravity and allows the airplane to fly in equilibrium. The lift force can be explained using Newton's Third Law.

Newtons's Third Law states that for every force exerted by one object on a second object (called the action force) there is an equal and opposite force exerted by the second object back on the first object (called the reaction force). In the case of an airplane wing, or airfoil, the action force is exerted by the wing on the air, and the reaction force is exerted by the air back on the wing, as shown in Figure 1. The component of this reaction force that is at right angles to the direction of motion is the lift force.

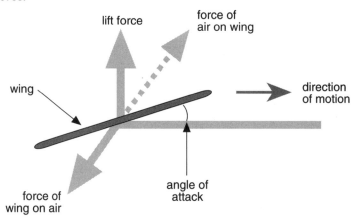

Figure 1: The lift force is at right angles to the direction of motion.

To generate lift, the wings of a flat-wing airplane like the Delta Dart must have an angle of attack, or upward tilt, as indicated in Figure 1. To create this angle of attack in the Delta Dart, the fuselage is tapered to the rear. To experience the effect of the wing tilt on lift force, hold your hand in a fast-moving airstream, first in a level position and then in a tipped-upward position. Like your tipped hand, the airplane wing forces or deflects the air downward and backward, and the reaction force of the air on the wing is forward and upward. The upward component of this reaction force is the lift on the wing. In most airplanes, cambered wings are used rather than flat wings. Cambered wings deflect the air downward more efficiently and create a larger lift force. A cambered wing does not have to be tilted; it creates lift because of its asymmetric shape.

The propeller also generates lift by acting as an airfoil moving in a vertical plane. The propeller flings the air backward, and the reaction force propels the plane forward. This type of lift is called thrust because the reaction force carries the plane forward, not up. This thrust, or "forward lift," is at right angles to the direction of the motion of the propeller.

As we have discussed, the airfoils (propeller and wings) create lift and enable the plane to move upward and forward. For a stable flight, the center of lift, or the point where lift forces can be considered to act, must be located at approximately the same point as the center of gravity. Otherwise, the lift forces may rotate the airplane out of control. Often, weights are placed on airplanes in order to move the center of gravity closer to the center of lift. Toy airplanes often have weights added near the nose of the airplane for this reason.

Control of an airplane's movement in flight is achieved by using control surfaces (rudder, elevator, and ailerons) to turn the airplane about three mutually perpendicular axes: horizontal lengthwise, horizontal crosswise, and vertical. Roll is banking or leaning movement about a horizontal lengthwise axis, pitch is climbing or diving movement about a horizontal crosswise axis, and yaw is a left and right movement about a vertical axis.

The rudder is attached to the vertical stabilizer and controls left and right movement by rotating to the side to which the airplane is to be headed. The force of the air on the stabilizer rotates the airplane. The elevator is connected to the rear of the horizontal stabilizer, which operates in the same way in controlling the up-and-down movement of the airplane. The ailerons, the movable part of a wing, control banking. To roll the airplane, one aileron is raised, increasing the lift on that wing, while the other aileron is lowered, "spoiling" or reducing the lift on that wing. The airplane rolls or banks toward the side with less lift.

Wing configuration also affects the stability of an airplane, and several forms of dihedral and cathedral configurations are used. In a dihedral configuration the wings are inclined upward (looking at the airplane head on), and in a cathedral configuration the wings are inclined downward. Examples of cathedral configurations are the B-52 bomber and the C-5A transport aircraft, with their large, downward-sloped wings. The Delta Dart uses a dihedral configuration to increase roll stability by introducing a constant self-adjusting tendency. When one wing lifts during flight, the other wing becomes more horizontal. The more horizontal wing produces more lift and moves upward, thus rolling the airplane towards stability.

Language arts:
- Have students write articles about their Delta Dart experiments for the school newspaper.

Math:
- Discuss the shapes of the different parts of the kit and dihedral angle.

Social studies:
- Have students research the history of flight and the social effects of aviation.

References

Teaching with Model Airplanes Programs (Teacher's Guide); Midwest Products: current edition.

Trinklein, F.E. *Modern Physics;* Holt, Rinehart and Winston: 1992.

Tropp, H.E.; Friedi, A.E. *Modern Physical Science;* Holt, Rinehart and Winston: 1991.

Contributor

Gary Lovely, Edgewood Middle School, Seven Mile, OH; Teaching Science with TOYS peer writer.

MINI MOTOR

Students use wire, a battery, and a magnet to make a simple electric motor.

Mini Motor

GRADE LEVELS

Science activity appropriate for grades 4–9
Cross-Curricular Integration intended for grades 7–9

KEY SCIENCE TOPICS

- current
- electric motor
- magnetism

STUDENT BACKGROUND

Students should have been introduced to the concepts of permanent magnets, batteries, and current.

KEY PROCESS SKILL

- investigating   Students investigate the effect of variables on the performance of their Mini Motors.

TIME REQUIRED

Setup	10	minutes
Performance	30	minutes
Cleanup	5	minutes

Materials

For the "Procedure"
Per student
- 1 D-cell battery (1.5 volts)
- 1 short length of ⅝-inch-diameter dowel rod
- 1 small piece of sandpaper
- tape
- 30 cm of 16- to 18-gauge steel or unvarnished copper wire
- 75 cm of 22- to 28-gauge varnished copper wire
- 1 doughnut-shaped ceramic magnet (2.8-cm-diameter)

 These magnets are available at Radio Shack stores.

Per class
- (optional) hot glue gun and glue
- several pairs of needle-nose pliers with cutting edges
- goggles

For the "Extension"
All materials listed for the "Procedure" plus the following:
Per class
- 6-volt or 9-volt battery
- square, oval, or rectangle-shaped objects
- magnets with varying strengths

Safety and Disposal

Wear goggles when cutting wire. No special disposal procedures are required.

Getting Ready

If time is limited or the students lack the ability to bend and cut the wire, the wire could be cut and even shaped before class.

Procedure

Have each student make a Mini Motor by doing the following:

1. Form the coil by wrapping 10 turns of 22- to 28-gauge varnished copper wire tightly around the dowel rod. Leave approximately 3 cm of wire on each end of the coil. (These will be the lead wires.) (See Figure 1.)

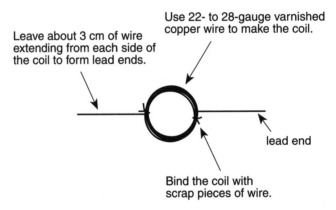

Leave about 3 cm of wire extending from each side of the coil to form lead ends.

Use 22- to 28-gauge varnished copper wire to make the coil.

lead end

Bind the coil with scrap pieces of wire.

Figure 1: Form the coil.

2. Use scrap wire to bind the coil wires to prevent unwrapping.

3. Straighten the lead wires, then sand the varnish off the leads.

   As an alternative, students can sand the varnish off one side of each lead (the top, for example). See the "Explanation" for a discussion of these two methods.

4. Cut a piece of bare copper or steel 16- to 18-gauge wire about 18 cm long.

5. Form a loop by wrapping one end of the wire about ¾ of a turn around a ⅝-inch dowel to form the contact for the battery's negative terminal. (See Figure 2.)

Teaching Physics with TOYS

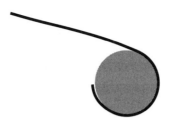

Figure 2: Wrap the end of the wire ¾ of a turn around the dowel.

6. Place the loop end of the wire under the D-cell battery and bend the wire upward 90° to fit next to the side of the battery. Tape the wire to the side of the battery. (This wire is the negative terminal lead.) (See Figure 3.)

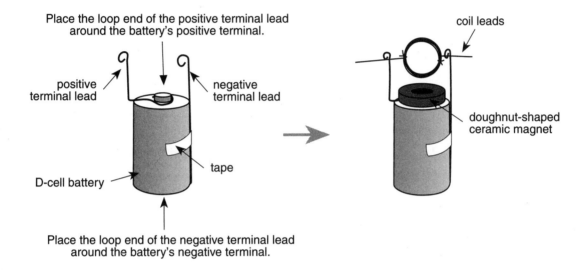

Place the loop end of the positive terminal lead around the battery's positive terminal.

coil leads

positive terminal lead

negative terminal lead

doughnut-shaped ceramic magnet

D-cell battery

tape

Place the loop end of the negative terminal lead around the battery's negative terminal.

Figure 3: Form the terminals, then place the magnet on top of the battery. Place the coil leads in the terminal loops.

7. Cut a piece of 16- to 18-gauge bare copper or steel wire approximately 12 cm long.

8. Form a loop on one end as in Step 5.

9. Place the loop end of the wire over the battery's positive terminal. (This wire is the positive terminal lead.)

10. Place the magnet over the wire loop on top of the battery. Bend the positive terminal lead up 90° to be parallel with the negative terminal lead.

11. Approximately 2.5 cm or more above the top of the battery, bend the free ends of the terminal leads into almost closed loops that are level with each other. (These will hold the coil leads.)

12. Place the coil leads into the open loops. Adjust the height of the loops until the coil is level and rests just above the circular magnet. Trim any excess wire. (See Figure 3.)

13. If necessary, push the coil gently to start it spinning. Observe the spinning coil. If it does not turn smoothly on the supports, adjust by bending the coil leads.

Extension

Have students change one or more of the following variables and observe the effect on the speed at which the coil spins:

- Change the energy source to a 6-volt or 9-volt battery.
- Make the coil larger or smaller.
- Change the shape of the coil to square, oval, or rectangular.
- Vary the number of turns in the coil.
- Vary the strength of the magnet used.

Explanation

 The following explanation is intended for the teacher's information. Modify the explanation for students as required.

In this activity, you make a simple direct current (dc) motor. The electric motor is a convenient source of motive power because it is clean and silent, starts instantly, and can be built large enough to drive a fast train or small enough to run a tiny wristwatch. Commercial motors may run on either direct or alternating current, may use permanent magnets or electromagnets, and may contain several coils. An electric motor requires three elements: a source of current (here provided as a direct current by the battery's potential), a magnet (here provided by a permanent ceramic magnet), and one or more loops of wire that are free to turn within the magnetic field of the magnet (here provided by the copper coil). An electric motor works by changing the chemical potential energy of the battery into mechanical energy in the form of a rotating coil.

When the circuit is closed (that is, when the negative and positive leads from the battery come into contact with the leads from the coil) the battery's potential produces a current in the copper wire. A loop carrying a current in a magnetic field has forces on it that cause it to turn, as shown in Figure 5. If the current flows in one direction without interruption, the force deflects the coil in one direction and it stays in that position. In essence, this is a galvanometer—an instrument to detect electric current.

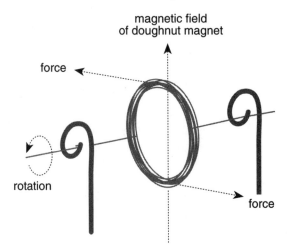

Figure 5: A loop carrying a current in a magnetic field
has forces on it that cause it to turn.

So how do you get from a galvanometer to a motor? The major difference between them is as follows: In a motor, the current does not continue to flow through the coil in one direction. Rather, each half-turn of the coil reverses the orientation of the coil leads with respect to the negative and positive terminal leads. This causes the flow of the current through the coil to be reversed with each half-turn. As a result, the force keeps changing direction, pushing the coil forward with every half-turn.

The Mini Motor in this activity works slightly differently than the motor just described. In the Mini Motor, the flow of the current through the coil is interrupted with each half-turn instead of reversing direction. When the current is flowing, the force deflects the coil, as in a galvanometer. When the current stops, the force stops pushing the coil, but the coil's inertia carries it through the turn and back to the starting position. When the current comes back on, the coil gets another push. A motor that operates in this way does not have as much force as the one with a reversing current because the coil is pushed only once with every full turn rather than once with every half-turn.

In the Mini Motor, the current can be interrupted in two ways. One method is to remove the varnish from only one side of each coil lead (for example, the top). With every half turn of the coil, the varnished side touches the battery leads, breaking the current. However, in this activity we removed the varnish from the entire coil lead. At first glance it seems that the motor should not work, because the current can flow through the coil leads without interruption no matter what the orientation of the coil. But because the coil bounces as it turns, the coil leads move away from the battery leads and break the connection. If the coil were perfectly stable and the coil leads remained in contact with the battery leads at all times, our method would not work.

Cross-Curricular Integration

Math:
- Have students calculate how much power they use to perform an athletic event, such as running up stairs carrying a weight. Power is work divided by the time interval over which the work is done. In the case of running up stairs with a weight, work is calculated by multiplying the total weight of the person plus weights times the height of the stairs. The time interval is the total time required to run up the stairs.

Social studies:
- Have students research and write a report on the history of electric motors or the impact of the invention of the electric motor on society.

Reference

Macaulay, David. *The Way Things Work;* Houghton Mifflin: Boston, 1988.
Renner, A.G. *How to Make and Use Electric Motors;* Putnam: New York, 1974.

Contributor

Gary Lovely, Edgewood Middle School, Seven Mile, OH; Teaching Science with TOYS peer mentor.

SOUND TUBE

Students use resonance to calculate the speed of sound.

Tuning fork and a water column

GRADE LEVELS

Science activity appropriate for grades 7–9
Cross-Curricular Integration intended for grades 7–9

KEY SCIENCE TOPICS

- amplitude
- frequency and natural frequency
- resonance
- standing waves
- vibration
- wavelength

STUDENT BACKGROUND

Students should have a basic understanding of waves, especially sound waves, and the wave parameters of frequency, wavelength, amplitude, and speed.

KEY PROCESS SKILLS

- interpreting data — Students calculate the speed of sound in air using resonance data.

- investigating — Students investigate resonance using tuning forks and air columns.

TIME REQUIRED

Setup	10	minutes
Performance	30	minutes
Cleanup	5	minutes

Materials

For the "Procedure"
Per class
- several objects, such as a block of wood and a metal pipe
- pendulum

> *A simple pendulum can be made by suspending an ordinary steel nut from the end of a string.*

Per student or group
- tuning fork

> *Any frequency of tuning fork can be used. This activity has been tested with 512 Hz.*

- tall graduated cylinder or other container in which to submerge at least 1 foot of the tube
- metric ruler
- plastic tube open at both ends, approximately 2.5 cm in diameter and at least 1 foot long

 The tubes are available from Consolidated Plastics Company, 1864 Enterprise Parkway, Twinsburg, OH 44087; (800) 362-3115.

For "Variations and Extensions"

❶ All materials listed for the "Procedure" plus the following:
Per class
- tuning forks of different frequencies

❷ Per class
- seashells of different shapes in which roaring sounds can be heard

❸ Per class
- access to a swing set

❹ Per class
- 512-Hz tuning fork
- paper roll tubes approximately 6 inches and 12 inches long

Safety and Disposal

No special safety or disposal procedures are required.

Introducing the Activity

Point out that it takes time for sound to travel a given distance. When compared to light this is very obvious. Remind the students that at a fireworks display, the burst of light is seen first and afterwards the sound is heard. This is because the speed of sound is much less than the speed of light.

Procedure

Part A: Demonstrating the Natural Frequency

1. From the back of the room, drop several objects one at a time. Have students attempt to identify the dropped objects without looking. Remind students that an object vibrates when it hits the floor and that the vibration is different for each object. This vibration is called the natural frequency of the object.

2. Explain that the sound heard as each object fell was a result of the vibrating object bumping into nearby gas molecules in the air and causing longitudinal waves to travel throughout the room. Tell students how we perceive sounds. (See the "Explanation.")

3. Show students the pendulum and start it swinging. Explain that the pendulum system has a natural frequency of vibration, just as the dropped object did. Point out that in this case the vibration (period) of the pendulum is easy to see.

4. Ask students if they can hear the pendulum swinging. *No.* Explain that the pendulum is colliding with gas molecules from the air, just as the dropped objects did, but the frequency is so low that we cannot perceive it. The vibration needed to create an audible (to humans) sound wave has to have a rate of more than 40 vibrations per second.

5. Ask students what they predict will happen if you strike the swinging pendulum. Purposely strike the pendulum at a faster, then slower rate than the natural frequency of the system. Then strike the pendulum at the same rate as the natural frequency of the system. Ask, "What happens?" *The pendulum swings higher.* State that this is an example of resonance. If desired, give other examples of resonance as described in the "Explanation."

6. Define the term "resonance" and explain that students will be using this phenomenon to calculate the speed of sound.

Part B: Calculating the Speed of Sound

Have each student or group do the following as instructed on the Data Sheet (provided):

1. Fill a tall graduated cylinder with water.

2. Hold the plastic tube in the water.

3. Strike the tuning fork and hold it over the top of the plastic tube. (See Figure 1.)

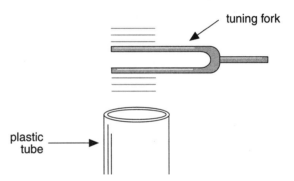

Figure 1: Hold the tuning fork over the top of the plastic tube.

4. Hold the tube all the way down in the water, then slowly raise the tube until you hear the water column start to "sound off" (resonate with the same pitch as the tuning fork).
The sound should be quite loud and distinct.

5. Measure the height of the plastic tube sticking out of the water and record.

6. Repeat Steps 3–5 two more times.

7. Use data from three trials to calculate the average tube height at which resonance is heard. Record data (average tube height, frequency of tuning fork, and air temperature) and calculate the speed of sound as shown on the Data Sheet.

Variations and Extensions

1. The students could first calculate the height of the air column required to match the frequency of the tuning fork (See the "Explanation"), then perform the test to check their predictions. Repeat with tuning forks of different frequencies.

2. Give students the opportunity to listen to the roaring sound inside several seashells of different shapes. Point out that the sound is caused by resonance, which is the amplification of one frequency determined by the shape of the seashell. This can also be shown by blowing across bottles filled with water to varying depths.

3. If a swing set is available, have students start swinging and then try pumping (or being pushed) faster than, slower than, and equal to the natural frequency of the swinging motion. Have them observe the effect of their pumping on the amplitude of the swinging motion.

4. As noted in the "Explanation," for a given fundamental frequency, resonance occurs for a tube open at both ends that is twice as long as a tube closed at one end. For a resonant frequency of 512 Hz, these lengths are approximately 6 inches for the tube closed at one end and 12 inches for the open tube. It turns out that two common cardboard paper roll tubes have approximately these lengths. Resonance for these two tubes can be demonstrated by laying the longer tube on its side with both ends open and standing the shorter tube vertically to close one end. The increase in the loudness of the sound as the tuning fork passes over one end of the tubes can be used to confirm the resonance.

Class Discussion

Use the phenomenon of lightning and thunder to reinforce this lesson. Discuss, using what students have learned, how to calculate the distance from a lightning bolt to your position. The flash of light is nearly instantaneous but the sound takes time to travel to the observer. The number of seconds between seeing the lightning bolt and hearing the thunder times the speed of sound (approximately 330 m/s) will give the distance to the lightning bolt.

Explanation

The following explanation is intended for the teacher's information. Modify the explanation for students as required.

All sounds are produced by the vibrations of material objects. As a vibrating object moves to and fro, it sends a disturbance through the surrounding medium, usually air, in the form of longitudinal waves. In a longitudinal wave, the medium vibrates in the same direction as the propagation of the wave. Sound waves consist of alternate regions of high and low pressure, which are known as compressions and rarefactions. As the object's surface moves forward in the air, it produces a compression. The surface then moves back, producing a rarefaction. Together each compression and rarefaction makes up a sound wave, and the waves move out in all directions. The stronger the vibrations of the object, the greater the pressure difference between the compressions and rarefactions, the larger the wave amplitude, and the louder the sound. The frequency of these compressions and rarefactions determines the frequency of the sound produced.

When sound waves reach our ears, our eardrums vibrate with the same frequency as the dropped object. These vibrations pass to the cochlea in the inner ear where they are converted to electric signals. The signals travel along the auditory nerve to the brain, and the sound is perceived. Point out that the frequency of vibration in the dropped object is too fast for us to see.

Resonance

When any object that can vibrate is disturbed, it will vibrate at its own special set of frequencies, which together form its special sound. These frequencies are called the object's natural frequencies, which depend on factors such as the material, size, and shape of the object. An object can have one or many natural frequencies. When an object is forced to vibrate at a natural frequency, the amplitude of vibration increases dramatically. This phenomenon is called resonance. A good way to think of resonance is to call it a sympathetic vibration. Literally, resonance means to resound, or sound again. One way to demonstrate resonance is to bring a vibrating object near another object with the same natural frequency and observe that the objects vibrate together.

Resonance does not just occur with sound waves; it occurs whenever successive forces are applied to a vibrating object in rhythm with its natural frequency. In "Introducing the Activity," you used a pendulum to demonstrate this idea. Another common experience illustrating resonance occurs on a swing (Extension 3). When pumping a swing, you pump in rhythm with the natural frequency of the swing. More important than the force with which you pump is the timing. Even small pumps or small pushes from someone else produce large amplitudes if delivered in rhythm with the natural frequency of the swinging motion. The increases in amplitude caused by resonance can sometimes cause problems. For example, English cavalry troops marching across a footbridge in 1831 inadvertently caused the bridge to collapse when they marched in rhythm with the bridge's natural frequency. Since then, it has been customary to order troops to "break step" when crossing bridges. In 1940, the Tacoma Narrows Bridge in the state of Washington was destroyed by wind-generated resonance.

Calculating the Speed of Sound

In the activity, we listen for resonance between the air column and the tuning fork. The natural frequency of the air column varies with its length. By raising the tube from the water, we slowly change the air column's natural frequency until it matches that of the tuning fork. When the tube reaches the proper position, a standing longitudinal wave is produced in the air column.

The basic expression for the velocity, v, of a wave in terms of its frequency, f, and wavelength, λ, is given by $v = f\lambda$. Thus, to calculate the speed of sound we need the resonant frequency (the frequency of the tuning fork) and the corresponding wavelength of the standing wave produced in the air column. Since we want to find the lowest natural frequency of the tube, called the fundamental frequency, we want the longest possible wavelength that can be sustained by the tube. For the fundamental, the length, L, of the air column is equal to ¼ wavelength of the sound wave, since the closed end (water) is a position of no motion of the air molecules and the open end (air) is a position where the air molecules have maximum motion, as shown below. The distance along a wave between zero and maximum amplitude is ¼ wavelength, therefore

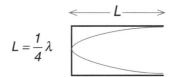

$$L = \frac{1}{4}\lambda$$

The speed of sound at 0°C is equal to the fundamental frequency times the wavelength and is equal to approximately 330 m/s. Increased temperature raises this speed slightly because faster-moving molecules in the warm air collide more often and can transmit a wave in less time. The correction is the addition of 0.6 m/s for every degree of temperature, T, above 0°C:

$$v = (300 + 0.6T)\,m/s$$

It is interesting to compare the fundamental frequencies of a tube closed at one end and a tube open at both ends. For the tube open at both ends, each end is a position of maximum motion of the air molecules; thus, there must be a position of zero motion at the center of the tube. Since the distance along a standing wave from maximum through minimum back to maximum amplitude is ½ wavelength, the length of the tube for the fundamental is ½ wavelength. This is to be compared to tube length equal to ¼ wavelength for a tube closed at one end. Thus, for a given fundamental frequency, resonance occurs for a tube open at both ends that is twice as long as a tube closed at one end.

For example, for a frequency of 512 Hz and using a speed of 330 m/s, resonance occurs for a tube closed at one end of length given by

$$L = \frac{\lambda}{4} = \frac{v}{4f} = \frac{330}{(4)(512)} = 0.161\,m = 6.3\,in$$

The length of a tube open at both ends for the same resonant frequency is twice as long, or approximately 13 inches.

Cross-Curricular Integration

Life science:
- Have students research the ranges of sound frequencies that various animals (especially dogs, bats, and whales) can perceive. Which animals can hear sounds that humans can't? How is the perception of these sounds important in the lives of each type of animal?

References

McAlexander, J.A. *Experiments for Technical Physics;* Allyn and Bacon: Boston, 1974; pp 243–246.

Trinklein, F.E. *Modern Physics;* Holt, Rinehart and Winston: Austin, TX, 1992.

Contributor

Gary Lovely, Edgewood Middle School, Seven Mile, OH; Teaching Science with TOYS peer mentor.

Handout Master

A master for the following handout is provided:
- Data Sheet

Copy as needed for classroom use.

Names
_____   _____

_____   _____

SOUND TUBE
Data Sheet

Objective

To calculate the speed of sound in air using a resonance tube.

Procedure

1. Calculate the anticipated speed of sound at room temperature:

$$330 \ m/s + temp. \, °C \times 0.6 \ m/s = v_o$$

2. Fill the graduated cylinder to the top with water.

3. Place the plastic tube all the way down into the water.

4. Strike the tuning fork and hold the tuning fork over the tube.

5. Slowly raise the tube.

6. When resonance occurs, stop immediately.

7. Measure the distance from top to level of water in centimeters.

8. Calculate the speed of sound with this formula:

$$v = f\lambda$$

Where: v = velocity
f = frequency
λ = wavelength

Data

f = Frequency of vibration of tuning fork = _____ Hz

$R.P.$ = Measured resonance point (average value) = _____ cm

λ = Wavelength = R.P. x 4 = _____ cm

v = Speed of sound (calculated) = _____ m/sec

v_0 = Speed of sound (anticipated) = _____ m/sec

BULL ROARER

*Students observe that the Bull Roarer converts the mechanical energy
of the rotating wood paddle into the energy of the roaring sound.*

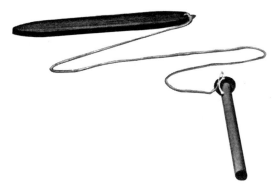

Bull Roarer

GRADE LEVELS

Science activity appropriate for grades 7–9
Cross-Curricular Integration intended for grades 7–9

KEY SCIENCE TOPICS

- aerodynamic lift
- frequency
- kinetic energy
- pitch
- potential energy
- rotational motion
- sound

STUDENT BACKGROUND

The students should have been introduced to the concepts of
rotational motion and energy and to the idea that vibration
causes compressions (condensations) and expansions
(rarefactions) that produce sound.

KEY PROCESS SKILL

- observing

Students observe the conversion of
mechanical energy to sound energy
as the toy is twirled.

TIME REQUIRED

Setup	5	minutes
Performance	15	minutes
Cleanup		none

Materials

For "Getting Ready" only

*These materials, intended for teacher use only, are needed to make the Bull Roarer the first
time the activity is done.*

- saw for cutting paddle
- drill and ⅛-inch drill bit

- slightly larger drill bit
- sandpaper

For the "Procedure"
Per class
- 1 Bull Roarer made from the following:
   - 1 piece of wood lattice strip, approximately ¼ inch thick x 1⅝ inches wide x 11½ inches long
   - 1 wooden dowel, ¾ inch x 9 inches
   - 1 #10 x 1¼-inch fender washer (large washer with small hole in center)
   - 1 #10 roundhead wood screw, 1 inch long
   - 1 twisted cord approximately 1 m long (from ball of white cord)

For "Variations and Extensions"
❷ All materials listed for the "Procedure" plus the following:
- 1 or 2 paddles of different dimensions

❹ All materials listed for the "Procedure" except
- substitute several flat pieces of cardboard and glue for the wood lattice strip

Safety and Disposal

Extreme caution must be observed when swinging the Bull Roarer so that no one is hit. The cord must be tied in bow-line knots as shown in Figure 1. No special disposal procedures are required.

Getting Ready

Construct the Bull Roarer:
Trim the corners of the wood lattice strip as shown in Figure 1. This will serve as the paddle. Drill a ⅛-inch-diameter hole for the cord as shown. Sand the edges enough to smooth off the splinters, but no more or the sound quality may be affected.

Drill a ⅛-inch-diameter hole in one end of the wooden dowel 1 inch deep. (See Figure 1.) Next, drill a clearance hole with a ³⁄₁₆-inch diameter about ⅜ inch deep in the same hole to clear the shank of the screw. This modified wooden dowel will serve as the handle.

Drill a ⅛-inch-diameter hole in the fender washer with its center approximately ⅛ inch from the outside of the washer. De-burr the hole by hand-twisting the tip of a larger drill bit in the hole or the cord will be cut. Screw the washer to the handle but do not tighten completely in order to allow the washer to rotate. Use bow-line knots to tie the cord to the small drilled hole in the washer and to the paddle. (See Figure 1.)

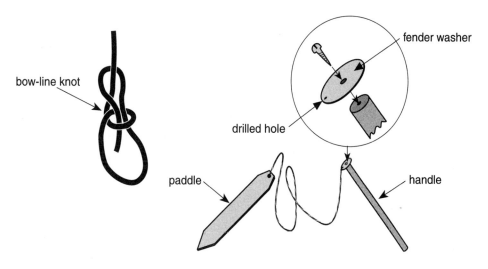

Figure 1: Assemble the Bull Roarer.

Procedure

1. Clear a large area in the room for the safe operation of the swinging Bull Roarer. Double check the knots to make sure the cord is secure.

2. Put a slight twist in the cord by turning the paddle while holding the handle stationary.

3. Hold your arm high over your head and swing the Bull Roarer in a horizontal circular pattern. The swing should be fairly fast and at a constant rate. Instruct students to carefully observe the movement of the Bull Roarer.

4. Have students discuss what they observed with a partner.

5. Demonstrate the Bull Roarer again if necessary to clear up any questions the students might have about what they observed in Step 3.

6. Have the pairs of students write down their observations.

7. Have a classroom discussion of the energy, sound, and frequency aspects of the Bull Roarer.

Variations and Extensions

1. Change the length of the cord and have students observe what changes occur in the movement of the Bull Roarer. This should affect the time it takes the Bull Roarer to switch between an "upward cone" rotation and a "downward cone" rotation. (See the "Explanation.")

2. Make a second or third paddle with different dimensions and test for pitch changes.

3. Calculate rotational rates from time and cord length.

4. Make a Bull Roarer from several thicknesses of cardboard glued together and kept flat.

Explanation

 The following explanation is intended for the teacher's information. Modify the explanation for students as required.

As the Bull Roarer is swung in a horizontal circle, the work done on the paddle and cord system goes into the kinetic energy of the system as it rotates in the circle and twists about an axis parallel to the cord. Kinetic energy is alternately stored as potential energy and recovered as the cord twists and untwists. The twisting motion slows down, stops, and reverses direction as the untwisting motion begins, and the cycle repeats as the system continues to move in a circle.

As the Bull Roarer moves and twists, some of the rotational kinetic energy is converted into sound energy and a roaring or whirring sound is heard. This results from the paddle beating against the air during its twisting motion and producing the condensation and rarefactions of a sound wave in air. The constant variation of the frequency of the twisting and untwisting motions causes the sound to have a varying frequency that is sensed as a changing pitch.

Although the system is swung in a horizontal circle, the motion of the paddle and cord is not horizontal but cycles between an upward cone and a downward cone. This behavior results from the upward and downward aerodynamic lift forces on the paddle as it twists in one direction and untwists in the opposite direction. The lift force is the component of the force of the air on the paddle that is at right angles to the horizontal motion of the paddle. The upward or downward direction of this force is determined by the direction of the twist motion of the leading edge of the paddle. This is similar to the upward and downward forces on horizontally thrown spinning baseballs. For both the paddle and baseball, the direction of the lift force is the direction in which leading edge moves because of spin or twist. The lift force is upward on a backward-spinning fastball and downward on an overhand curveball thrown with topspin.

Cross-Curricular Integration

Life science:
- Discuss the production of sound by flying animals and compare how flying animals make sound to how the Bull Roarer makes sound. Discuss possible reasons why some animals make sound in flight (such as cicadas) while others are silent (such as owls).

Social studies:
- Have students research and write reports about how the Bull Roarer was used as a folk toy and to communicate in different societies.

Reference

Schnacke, D. *American Folk Toys;* Putnam: New York, 1973.

Contributor

Gary Lovely, Edgewood Middle School, Seven Mile, OH; Teaching Science with TOYS peer mentor.

SINGING CHIMES

Students are motivated by music to explore sound.

Set of chimes

KEY SCIENCE TOPICS

- natural frequency
- sound
- pitch
- vibration

STUDENT BACKGROUND

Students should be familiar with the concepts of frequency, pitch, and sound waves.

KEY PROCESS SKILLS

- observing — Students observe sounds produced by pipes of various lengths.

- hypothesizing — Students hypothesize how changes in length of pipe will affect pitch.

TIME REQUIRED

Setup	10	minutes (plus 1 hour first time)
Performance	25	minutes
Cleanup	5	minutes

Materials

For "Getting Ready" only

These materials, intended for teacher use only, are needed to make the chimes the first time the activity is done.

- pipe cutter, such as General #120
- metric ruler
- (optional) electric drill with ⅛-inch drill bit
- slightly larger drill bit
- file
- goggles

For the "Procedure"
Per class

- 1 set of chimes (made from 2 10-foot pieces of ½-inch galvanized electrical conduit)

  Have these pre-cut into 5-foot lengths at the hardware store for handling ease.

- 1 of the following:
  - spool of string or monofilament fishing line
  - 40 #10 or similar-sized rubber bands
- 20 spoons or large nails
- permanent marker

For "Variations and Extensions"
❶ Per class
- several objects, such as a fork, a metal wrench, and wooden spoon or rod

❷ Per class
- several glass soft-drink bottles or other narrow-necked glass bottles

❹ All materials listed for the "Procedure" plus the following:
Per class
- materials to build more than 1 set of chimes

Safety and Disposal

Goggles should be worn when cutting or drilling. Materials should be properly clamped before cutting or drilling. No special disposal procedures are required.

Getting Ready

Construct chimes:
Make the 20 chimes, numbered 1–20 from the four 150-cm (5-foot) pieces of conduit pipe. Cut the chimes from these pieces of pipe as follows:

Table 1: Chime Lengths							
Cut From Pipe 1		Cut From Pipe 2		Cut From Pipe 3		Cut From Pipe 4	
Chime #	Length (cm)	Chime #	Length (cm)	Chime #	Length (cm)	Chime #	Length (cm)
1	38.5	5	34.2	8	31.1	13	27.2
2	37.5	6	33.1	9	30.3	14	26.3
3	36.2	7	32.2	10	29.7	15	25.5
4	35.1	17	23.8	11	28.9	16	24.7
		18	23.1	12	28.1	19	22.5
						20	21.8

Mark and cut the chimes one at a time to avoid measuring errors associated with the width of the cut. Cut the chimes with a pipe cutter to give smooth edges. File any sharp edges.

If the chimes are to be played on desks, place one rubber band approximately 1½ inches from each end. The positions of the rubber bands can be adjusted slightly to give the best sound from the chimes. If the chimes are to be played while being held vertically, drill a ⅛-inch hole approximately 1 inch from the top of each chime. De-burr the drilled holes by hand twisting the tip of a larger drill bit in the holes. Insert string or monofilament line through each hole and tie it off to form a 3- to 6-inch loop. Mark the number and note on each chime with a permanent marker. (Refer to Table 2.)

Make an overhead transparency of the Song Sheet (provided) or write the numbers of the chimes for each song on the board.

Procedure

1. Ask for a student volunteer to play each chime required in the songs that are to be performed; give each a spoon or nail to use as a striking rod.

2. Using the songs on the Song Sheet (provided), write the numbers of the chimes to be played on the overhead projector or the board. Explain that when you point to a number, the student with that number will strike his or her chime.

3. Have the students perform the song as you point to the numbers in sequence.

4. When this activity has reached some degree of success, take time to question why the different chimes have different sounds. Guide the students to discover that the length of a chime is related to the pitch of the sound they hear from the chime.

Variations and Extensions

1. Out of the students' range of sight, drop several objects one at a time, such as a metal fork or wrench and a wooden spoon or rod. Have the students guess what you dropped. When they successfully identify the objects, explain that they were able to do so because they recognized the natural frequencies of these items.

2. Fill identical glass bottles with different amounts of water. Strike these containers to show the students that each has a different natural frequency. This frequency deals with vibrating the bottle and the water. Next, blow across the tops of the bottles. There should be a similar variation in sounds, in this case due to the length of the air column in each bottle. Have students create a tune by experimenting with varying amounts of water in a series of bottles.

3. Transcribe simple songs by converting notes into the corresponding numbers on the chimes.

4. Make more than one set of chimes to allow all students to participate at once.

Explanation

The following explanation is intended for the teacher's information. Modify the explanation for students as required.

A collection of sounds is called music when the individual sounds are pleasing and the sounds are arranged into pleasing patterns. Of course, our human definition of "pleasing" varies widely with culture and experience.

Although music is aesthetically pleasing, from the point of view of physics the sounds we call music are no different than other vibrations. All sounds are produced by the vibrations of material objects. As a vibrating object moves back and forth, it sends a disturbance through the surrounding medium, usually air, in the form of longitudinal waves. In a longitudinal wave, the direction of the disturbance is the same direction as the propagation of the wave. The waves consist of alternate regions of high and low pressure, which are known as compressions and rarefactions. As the object's surface moves forward in the air, it produces a compression. The surface then moves back, producing a rarefaction. Together each compression and rarefaction makes up a sound wave, and the waves move out in all directions. Usually, the frequency of the vibrating source and the frequency of the sound waves produced are the same.

The more rapidly an object vibrates, the greater the number of waves per unit time (frequency). The unit of frequency is called hertz (Hz), after Heinrich Hertz who demonstrated radio waves in 1886. One vibration per second is one hertz. The stronger the vibrations of the object, the greater the pressure difference between the compressions and rarefactions and the louder the sound. The subjective or sensory description of vibration frequency is known as pitch.

When any object that can vibrate is disturbed, it will vibrate at its own special set of frequencies, which together form its special sound. These frequencies are called the object's natural frequencies, and they depend on factors such as the material, elasticity, and shape of the object.

The pipes used in this activity make good musical instruments because their natural frequencies produce pleasing sounds. Each pipe is cut at a particular length so that it will vibrate with a particular natural frequency. The longer the pipe, the slower it vibrates; the shorter the pipe, the faster it vibrates. (See Table 2.) When speaking of music, each frequency is referred to as a note. Humans perceive frequency as pitch. Although pitch depends primarily on frequency, it also is influenced by other quantities such as intensity.

Table 2: Chime Frequencies

Chime #	Note	Frequency (Hz)	Chime #	Note	Frequency (Hz)	Chime #	Note	Frequency (Hz)
1	F	175	8	C	262	15	G	392
2	F (sharp)	185	9	C (sharp)	277	16	G (sharp)	415
3	G	196	10	D	294	17	A*	440
4	G (sharp)	208	11	D (sharp)	311	18	A (sharp)	466
5	A	220	12	E	330	19	B	494
6	A (sharp)	233	13	F	349	20	C	523
7	B	247	14	F (sharp)	370			

* This is A_4=440Hz.

Cross-Curricular Integration

Language arts:
- Have students research and write reports on the importance of music in different societies and different eras.

Math:
- As a class, compare the length of each pipe to its corresponding frequency and graph these findings to see what relationship exists.

Music:
- Have students compare notes of the singing chimes to notes played on a variety of instruments. Students could also compose some original music to be played on the singing chimes.
- Have students investigate the types of musical instruments used in cultures around the world and try to determine how the player of each type of instrument varies the frequencies produced so that he or she can play music.
- As a class, study and play music of different countries on the singing chimes.

Reference

McGrath, S. *Fun with Physics;* National Geographic Society: Washington, DC, 1986.

Contributors

Lynda Dunlap, Southwestern Elementary School, Patriot, OH; Teaching Science with TOYS, 1991–92.
Gary Lovely, Edgewood Middle School, Seven Mile, OH; Teaching Science with TOYS peer mentor.

Handout Master

A master for the following handout is provided:
- Song Sheet

Copy as needed for classroom use.

SINGING CHIMES
Song Sheet

Michael Row the Boat Ashore

10	14	17	14	17	19	17
	10	14	10	14	15	14

14	17	19	17
10	14	15	14

14	17	17	14	15	14	12
10	14	14	10	12	10	9

10	12	14	5	12	10
7	9	10		9	

Happy Birthday

8	8	10	8	13	12
		6		8	6

8	8	10	8	15	13
		6		8	8

8	8	20	17	13	12	10
		17	13	8	8	6
		13	8			

18	18	17	13	15	13
13	13	13	8	12	8
10	10	8		8	

The Star-Spangled Banner

Ohh	—oh	say	can	you	see
9	6	2	6	9	14
				6	9
					6

by	the	dawn's	ear-	ly	light
18	16	14	6	8	9
9	7	6	2	4	4

what	so	proud-	ly	we	hailed
9	9	18	16	14	13
6	6	9	13	9	7

at	the	twi-	light's	last	gleam-	ing,
11	13	14	14	9	6	2
7	7	9	9	6		
				2		

[Continued.]

whose	broad	stripes	and	bright	stars
9	6	2	6	9	14
				6	9
					6

through	the	pe-	ri-	lous	fight
18	16	14	6	8	9
9	7	6	2	4	4

o'er	the	ram-	parts	we	watched
9	9	18	16	14	13
6	6	9	13	9	7

were	so	gal-	lant-	ly	stream-	ing?
11	13	14	14	9	6	2
7	7	9	9	6		
				2		

And	the	ro-	ckets'	red	glare
18	18	18	19	21	21
14	14	14	14	14	14

the	bombs	bur-	sting	in	air
19	18	16	18	19	19
14	14	13	14	13	13
	9	9	9	9	9

gave	proof	through	the	night
19	18	16	14	13
13	13	11	10	10
9	9			

that	our	flag	was	still	there.
11	13	14	6	8	9
7	9	9	2	2	4
	7	7			

Oh	say	does	tha-	at	star-	span-	gled
9	14	14	16	18	11	11	11
7	9	9	9	9	7	7	7
	6	6	6	6			

ban-	ne-	-er	ye-	et	wa-	ave
16	19	18	16	14	14	13
	14	14	14	9	9	7
			9			

o'er	the	la-	and	of	the	free
9	9	14	16	18	19	21
7	7	9	13	13	13	14
			9	9	9	9

and	the	home	of	the	brave?
14	16	18	19	16	14
9	9	13	13	13	9
7	8	9	9	9	6

TUBE SPECTROSCOPE

Students discover that white light is composed of different wavelengths that can be seen in an easy-to-make spectroscope.

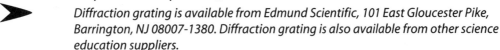

Tube Spectroscopes

Materials

For the "Procedure"
Per student

- 1 piece of plastic diffraction grating (thin plastic containing about 13,000 fine parallel lines per inch), about 1 cm x 1 cm

 Diffraction grating is available from Edmund Scientific, 101 East Gloucester Pike, Barrington, NJ 08007-1380. Diffraction grating is also available from other science education suppliers.

- cardboard tube

Toilet paper tubes are ideal. Have each student bring one from home.

- lightweight, opaque cardboard or card stock, such as black posterboard
- transparent tape or glue stick

- white craft glue or black electrical tape
- scissors
- utility knife with retractable blade
- colored pencils, markers, or crayons

For "Variations and Extensions"

❶ Per class
- Tube Spectroscope (made in the "Procedure")
- high-voltage power supply
- spectral gas tubes

❷ Per class
- film can for each student
- drill and ¼-inch drill bit

Safety and Disposal

Instruct students to always direct a sharp blade away from their bodies and other students. No special disposal procedures are required.

Procedure

Part A: Making the Spectroscope

Have each student do the following:

1. Draw a cover for one end of the toilet paper roll by tracing around one end of the toilet paper roll on the cardboard.

2. Add cardboard tabs to the cover by drawing three rectangles around the edge of the circle approximately as shown in Figure 1.

 Figure 1 is the correct size for the end covers and could be used as a pattern, although exact size and position of the tabs is not important.

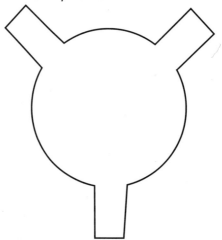

Figure 1: Leave tabs on the cardboard circles.

3. Repeat Steps 1 and 2.

4. Cut out the covers.

5. Cut a slit in one end cover and a square hole in the other as shown in Figure 2.

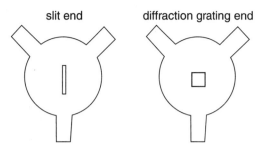

Figure 2: Prepare the cardboard covers as shown.

6. Position the diffraction grating over the square hole so that the grating covers the hole.

7. Carefully fix the grating in place with transparent tape or a glue stick.

The tape should not overlap into the opening for the diffraction grating.

8. Position the diffraction grating end cover over one end of the cardboard tube, fold down the three tabs, and tape this end to the tube.

9. Position the slit end on the other end of the cardboard tube. Before folding the tabs and taping, point the spectroscope toward a window or other light source and look through the spectroscope. Rotate the slit until the spectrum is spread to its maximum extent at right angles to the slit.

10. When the slit end is positioned correctly, fold down the three tabs, and tape. Seal the ends with black electric tape or craft glue to block out any light.

Part B: Making Observations

1. Have students use the spectroscope to view various light sources in the classroom (for example, fluorescent, incandescent, daylight near the sun, and any other light source available).

Be sure that students do not look directly at the sun.

2. Instruct students to make detailed observations of the spectrum they see and make a colored drawing of a spectral band.

Most classroom sources of white light provide a continuous spectrum.

3. Tell student to take their spectroscopes home and look for light sources that provide a bright line spectrum, and then draw the bright line spectra observed for each.

See the "Explanation" for suggested light sources.

Variations and Extensions

1. If you have a spectral gas tube with power supply, students can also see good bright line spectra with the spectroscopes.

2. Have students make spectroscopes another way by doing the following: use a utility knife to cut a slit in the bottom of a film can; drill a ¼-inch-diameter hole in the lid of the film can; tape a piece of diffraction grating over the hole on the inside of the lid. Rotate the lid until the observed spectrum is spread to its maximum extent at right angles to the slit.

3. Students can observe line spectra by flame-testing various materials. Kits for this purpose are available from Central Scientific Co., 11222 Melrose Avenue, Franklin Park, IL 60131-1364; (800) 262-3626.

Explanation

 The following explanation is intended for the teacher's information. Modify the explanation for students as required.

In this activity, you construct a simple version of a spectroscope, an instrument used to observe the individual wavelengths that make up a light source. One source of light is heated solids. When heated, solids emit light over a wide range of wavelengths. Many light sources behave this way, for example the filament of a tungsten light bulb and a metal object heated in a flame. When light sources emit all or nearly all visible wavelengths at once, we perceive the color of light as white. When white light passes through the multitude of closely spaced parallel slits that make up the diffraction grating of the spectroscope, an interference pattern is formed in which the light is separated into different wavelengths, or colors. This is called the continuous spectrum and resembles a rainbow ranging from red to violet.

Spectra that are discrete rather than continuous contain individual wavelengths, or colors, and are called line spectra. Line spectra are emitted by gases or other vaporized materials that have been energized with thermal energy or a high-voltage electrical discharge. The input of energy temporarily excites electrons in the material, boosting them to higher energy levels. As the electrons drop back to lower energy levels, they give up energy in the form of light. The difference in energy between the two levels determines the wavelength of light emitted. When we view the material through a spectroscope, we see a pattern of bright lines; thus the name line spectrum. Every element has a unique line spectrum. The characteristic patterns can be used to identify the atoms in an unknown sample, much as fingerprints are used to identify people.

Several types of lamps exploit the ability of a gas to emit light when excited by an electric discharge. Street lights commonly contain mercury or sodium vapor. The visible portion of mercury emission falls mainly in the yellow, green, and violet regions, and sodium emission is centered mainly in the yellow region. A fluorescent lamp is a tube of mercury vapor with the inner surface coated with a fluorescent material. The mercury itself emits a line spectrum, but when the light emitted by the mercury strikes the fluorescent material, the fluorescent material absorbs the light and then emits a wide range of wavelengths that combine to form white light. Thus, when you view fluorescent lights through the spectroscope, you see a continuous spectrum. Sometimes the mercury vapor line spectrum can be seen superimposed on the continuous spectrum. The "neon" signs seen on storefronts depend on light emission by excited gases, but do not always contain neon. Depending on the color desired, various gases can be used, including argon (blue-violet) and krypton (white). In addition, special fluorescent materials on the inner glass walls of the tubes produce greens and other colors.

Cross-Curricular Integration

Earth science:
- Students can investigate how spectroscopy helps us identify the elements that the sun, other stars, and the atmospheres of planets are made of without taking direct samples of their composition.

Language arts:
- Have students read myths from different cultures about rainbows and then research to find the science behind rainbows.

Social studies:
- Have students research the history of a variety of gas bulbs, such as neon and mercury vapor streetlights.

References

Chang, R. *Chemistry;* McGraw-Hill: New York, 1991

Fariel, R.F. *Earth Science;* Addison-Wesley: Menlo Park, CA, 1984.

Trinklein, F.E. *Modern Physics;* Holt, Rinehart and Winston: Austin, TX, 1992.

Contributors

Steven Hess, Verity Middle School, Middletown, OH; Teaching Science with TOYS, 1988–89.

Gary Lovely, Edgewood Middle School, Seven Mile, OH; Teaching Science with TOYS peer mentor.

Activities Indexed by Key Process Skills

Process Skill	Gravity Makes Things Fall	Comparing Mass Using a Pan Balance	Measuring Mass Using a Pan Balance	Ramps and Cars	Balloon on a String	Ping-Pong™ Puffer	Balancing Stick	The Skyhook	The Six–Cent Top	Bouncing Balls	Snap, Crackle, Pop	Magic Balloon	Magnet Cars	School Box Guitar
									Grades K–3 Activities					
1. Observing		●			●			●			●		●	
2. Communicating									●					
3. Estimating			●							●				
4. Measuring			●											
5. Collecting Data								●				●		
6. Classifying														
7. Inferring						●								
8. Predicting				●	●									
9. Making Models														
10. Interpreting Data														
11. Comparing/Contrasting														
12. Making Graphs										●				
13. Hypothesizing		●					●	●						
14. Controlling Variables				●										●
15. Defining Operationally	●											●		
16. Investigating	●						●				●	●		

Grades 4–6 Activities

Process Skill	Forces and Motion	Crash Test	Two-Dimensional Motion	Understanding Speed	Push-n-Go®	The Toy that Returns	Physics with a Darda® Coaster	Exploring Energy with an Explorer Gun®	Bounceability	Energy Toys Learning Center	Simple Machines with Lego®	Gear Up with a Lego® Heli-Tractor	Levitation Using Static Electricity	Doc Shock
1. Observing	●	●					●							
2. Communicating										●	●	●		
3. Estimating														
4. Measuring			●						●					
5. Collecting Data				●				●						
6. Classifying											●	●		
7. Inferring						●							●	
8. Predicting							●							
9. Making Models														
10. Interpreting Data														
11. Comparing/Contrasting														
12. Making Graphs				●				●	●					
13. Hypothesizing	●				●		●							●
14. Controlling Variables			●						●					
15. Defining Operationally														
16. Investigating		●			●	●				●				●

Process Skill	Walking Feet	Downhill Racer	The Projectile Car	Balance Toys and Center of Gravity	Whirling Stopper	Floating Cans	Falling Filters	Stick Around	Delta Dart	Mini Motor	Sound Tube	Bull Roarer	Singing Chimes	Tube Spectroscope
Grades 7–9 Activities														
1. Observing					●		●	●				●	●	●
2. Communicating														
3. Estimating														
4. Measuring		●	●		●									
5. Collecting Data	●						●							
6. Classifying														
7. Inferring														
8. Predicting						●			●					
9. Making Models														
10. Interpreting Data											●			
11. Comparing/Contrasting														
12. Making Graphs	●													
13. Hypothesizing													●	
14. Controlling Variables		●	●						●					
15. Defining Operationally														
16. Investigating				●		●			●	●	●			

Appendix B:
Activities Indexed by Topics

Appendix C:
Alphabetical Listing of Activities